PSYCHO-MYCOLOGICAL STUDIES

Number One

HALLUCINOGENIC PLANTS
OF NORTH AMERICA

Hallucinogenic Plants of North America

by Jonathan Ott

Wingbow Press

Berkeley

Published by *Wingbow Press.*
Distributed by *Bookpeople,*
2940 Seventh Street, Berkeley, California, 94710.

Library of Congress Catalog Card Number: 76-6216
ISBN: 0-914728-15-6 (paper), 0-914728-16-4 (cloth)

Designed by Hal Hershey
Illustrations by Mary Jo Eloheimo

Frontispiece: Pre-Columbian mushroom stone from highland Guatemala, circa 500 B.C.
Cover: Peyote cactus in flower.

The publisher gratefully acknowledges Mrs. Joy Spurr, the Puget Sound Mycological Society, and the following individuals for their cooperation in providing photographs for this book: Ray Bess, plate 24; Dr. Michael Beug, plate 9; Jeremy Bigwood, plates 1, 2, 6, 7, 8, 11, 12, 16, 17, 18, 19, 20, 21, 22, 23, 25, 27, 28; Jerry R. Boydston, plate 26; Dr. William S. Chilton, plate 4; Edward Costello, plates 14, 15; Robert Gerrish, plate 5; Joy Spurr, plates 10, 13; Dr. Daniel E. Stuntz, plate 3.

The publisher gratefully acknowledges the following for permission to quote:

From JUNKIE. Copyright © 1953 by William Burroughs. Published by Ace Books, New York. By permission of the author and publisher.

From SATANISM AND WITCHCRAFT by J. Michelet. Copyright © 1965 by A. Allinson (translation). Published by permission of Citadel Press (a division of Lyle Stuart, Inc.) 120 Enterprise Avenue, Secaucus, New Jersey, 07094.

From THE GUERMANTES WAY by Marcel Proust. Copyright © 1970 by C. Moncrieff (translation). Published by permission of Random House, Inc., New York.

From CEREMONIAL CHEMISTRY: The Ritual Persecution of Drugs, Addicts, and Pushers. Copyright © 1974 by Thomas Szasz. Published by Anchor Press/Doubleday, New York. By permission of the author.

From MARIA SABINA AND HER MAZATEC MUSHROOM VELADA. Copyright © 1974 by R. Gordon Wasson. Published by Harcourt, Brace, Jovanovich, New York. By permission of the author.

From SOMA: DIVINE MUSHROOM OF IMMORTALITY. Copyright © 1968 by R. Gordon Wasson. Published by Harcourt, Brace and World, New York. By permission of the author.

From MUSHROOMS, RUSSIA AND HISTORY. Copyright © 1957 by R. Gordon Wasson. Published by Pantheon Books, New York. By permission of the author.

From *The Hallucinogenic Mushrooms of Mexico: An Adventure in Ethnomycological Exploration.* Copyright © 1959 by the New York Academy of Sciences. Published in *TNYAS*, Series II, Vol. 21, #4, February, 1959. By permission of the author and the New York Academy of Sciences.

Revised Edition: October 1979

Dedicated to the Memory of

BLASIUS PAUL (BLAS PABLO) REKO
1876–1953

Pioneer Ethnobotanist in Mexico
First Scientist, with R. E. Schultes, to
Collect Identifiable Material of the
Aztec Magic Plants *Teonanácatl* and *Ololiuqui*

CONTENTS

INTRODUCTION

In many cultures around the world, hallucinogenic plants have long played vital roles in every aspect of living and dying. As a result of their extraordinary effects, they are held to be organisms inhabited by supernatural forces and, consequently, are sacred, sometimes even deified.

During the past century, these narcotics have attracted the attention of artistic, literary, and scientific thought in western cultures. In the last three or four decades especially, growing numbers of people in nearly all sectors of our society have experimented with some of the hallucinogens. Occasionally, this experimentation has apparently assumed almost epidemic proportions, and our society has become heatedly divided by the recent flirtation with these drug plants from exotic cultures or with active chemical substances derived from them. There seems to be little or no middle ground in the arguments that swirl around the use of hallucinogens in our midst.

Into such an environment this book is born. It comes as a fresh breath and with an understanding look at the historical development of hallucinogens in human affairs. Basically an anthology, it employs exquisitely chosen quotations from a wide spectrum of literature—both ancient and modern—and, often in truly poetic fashion, weaves around the selections the author's searching, sympathetic, meaningful, and, on occasion, even challenging interpretations. It rightly could be described as an anthological commentary on the hallucinogenic scene in human history.

The work of a very young man of great promise, *Hallucinogenic Plants of North America* will surely take its place as a book—and there are far too few of them—destined never to be outdated simply because it is built of the stuff that thinking man of any period of time needs if he is to be understanding of the past, sympathetic of the present, and inquisitive of the future.

Richard Evans Schultes
Professor of Biology
Director and Curator of Economic Botany
Botanical Museum of Harvard University
Cambridge, Massachusetts

PREFACE

The law which is good,
Lawyer woman am I,
Woman of paper work am I,
I go to the sky,
Woman who stops the world am I,
Legendary woman who cures am I.

—María Sabina,
quoted by R. G. Wasson in
María Sabina and her
Mazatec Mushroom Velada

The use of hallucinogenic drugs for recreation has become so widespread in recent years, that one commonly hears the term "drug culture" applied to that group of persons involved in this use. Of course, we all belong to a "drug culture;" the average American, young or old, will commonly ingest two or three biodynamic substances every day. The profound lack of awareness implicit in this habitual use, and the failure to identify these habits with "drug use" is little short of astonishing, considering the fact that the use of certain types of drugs, particularly the hallucinogens, is so widely publicized and so vigorously persecuted.

Although the users of hallucinogenic drugs, by virtue of the fact that these drugs have profound effects, are generally more aware of their drug use than are, say, coffee drinkers,* they are, as a rule, grossly misinformed as

---

*I am not being facetious. Who is "intoxicating" himself more—the coffee drinker, whose average cup of coffee contains 100–150 mg of the alkaloid caffeine, or the recreational user of hallucinogenic mushrooms, whose average dose of mushrooms contains 5–10 mg of the alkaloid psilocybin? Who takes more frequent doses?

to the nature and consequences of their drug experiences, and have little or no conception of the historical importance of the drugs they use. Myths and fallacies everywhere abound. Young drug users often complain that police, parents, teachers, and physicians seem to know little about hallucinogenic drugs, despite their positions of authority. This is, unfortunately, true—it is also apparent that these young drug users themselves know very little about the drugs they so commonly use.

The phenomenal popularity of the recent series of books by Carlos Castaneda (27), detailing his experiences with Don Juan, a now legendary Mexican shaman, has indicated how widespread is the current interest in drug plants and altered states of consciousness. Castaneda's books, although of unquestionable value as social documents, have served to cultivate a mystique surrounding the use of hallucinogenic plants. Having worked in Mexico, I have seen many people, from all over the world, visiting this country in hopes of finding Don Juan, María Sabina (188), or some other such person who can introduce them to these mysterious plants. Usually, this search is fruitless. Having spoken to many such persons, I have perceived that they have no clear conception of the object of their search. Indeed, if these people had known anything about hallucinogenic plants, I feel certain that they would have looked, literally, in their own back yards before coming to Sonora or Oaxaca. These plants grow all over the world.

I feel that this mystique surrounding the use of hallucinogenic plants is unfortunate and should be dispelled. Accordingly, I have decided to publish this information regarding the hallucinogenic plants used in Don Juan's sorcery. Don Juan calls these plants: "devil's weed," or "toloache" (plants of the genus *Datura*), "Mescalito" (the peyote cactus, *Lophophora williamsii*), and "honguillos" (hallucinogenic mushrooms, presumably of the genus *Psilocybe,* used in "humito," or "the little smoke") (27). Additional information is included on several other hallucinogenic plants which commonly grow in North America.

Many of the plants discussed in this book are, by virtue of their chemical constitution, illegal under federal law. Under the provisions of Public Law 91-513, Section 202, Schedule 1(c) and Schedule 3(b), unauthorized possession, sale, or use of certain hallucinogenic compounds, or any materials containing these compounds, is declared illegal. Accordingly, the following plants discussed in this book are proscribed: all mushrooms of the genus *Psilocybe* and the genus *Panaeolus; Lophophora williamsii;* all plants of the genus *Cannabis;* and all plants of the family *Convolvulaceae.* The remaining plants are, at the time of this writing (5/75), not subject to federal controls, but may be subject to state or local regulation.

Prospective users are advised to exercise extreme caution. With respect to the fungi, it would be wise to consult other sources, to become more familiar with the various types of toxic mushrooms. Orson Miller's *Mushrooms of North America* (121) has excellent photographs and taxonomic information, although some of the information on toxicity is highly inaccurate. Both Miller's book, and the present volume contain photographs of two deadly poisonous mushrooms, *Amanita phalloides* and *Galerina autumnalis*. I feel that these two species represent the only potentially lethal types which could conceivably be mistaken for hallucinogenic mushrooms. Under the photographs of these two species is presented information enabling the reader to distinguish *A. phalloides* from the hallucinogenic *Amanita* species, and *G. autumnalis* from the various species of *Psilocybe* to which it bears a superficial resemblance. There are other species of potentially lethal mushrooms growing in North America, notably *Amanita virosa, A. verna, A. tenuifolia, A. bisporigera, Galerina venenata, G. marginata,* and *Conocybe filaris*—anyone collecting wild mushrooms with the intention of eating them should become familiar with these species.

I do not recommend ingesting plants of the genus *Datura* or the genus *Atropa*. These plants are very toxic and may produce unpleasant effects, even death.* Should the reader decide to try these plants, he should start with extremely small amounts, and gradually increase this quantity, until he arrives at a safe dose level. It is ironic that these plants are perfectly legal under Title 21 U.S. Code, while several far more innocuous plants are classified as illegal.

This book has been designed for use as a field manual and reference guide to the literature on the chemistry and history of hallucinogenic plants. The first part contains illustrations and descriptions of the botany, chemistry, and history of 30 representative hallucinogenic plants of North America. These plants are grouped by families, in alphabetical order. References are included to the chemical, historical, and botanical literature. The second part presents a theory, based on historical evidence, concerning the etiology of primitive religions in hallucinogen cults, and the consequent evolution of these religions into modern types. Dealing as it does with antiquity, this section is of necessity highly speculative. The third part of the book discusses the biochemistry and physiology of the

---

*It has recently been reported that a number of deaths in Kern County, California have been attributed to overdoses of *Datura innoxia* by would-be recreational users (Robert Bye, personal communication).

central nervous system, and presents models for the induction of hallu-cinosis by both chemical and non-chemical means. This section is more technical than the rest of the book, and probably requires some prior familiarity, on the part of the reader, with introductory level biology and chemistry. Included in the book are three appendices: the first is designed to show structural relationships between hallucinogens and neurochemi-cals, the second is a glossary, to assist the reader with technical termino-logy, and the third is a description of simple field and laboratory tech-niques to aid the amateur in the collection and identification of specimens. Finally, a list of general interest books relating to the study of hallucino-gens and altered states of consciousness is presented, as well as an extensive bibliography, giving the references to the scientific and historical literature on hallucinogenic plants.

FOREWORD

*Not far thence is the secret garden in which grow like strange
flowers the kinds of sleep, so different one from the other, the sleep
induced by datura, by the multiple extracts of ether, the sleep of
belladonna, of opium, of valerian, flowers whose petals remain
shut until the day when the predestined visitor shall come and,
touching them, bid them open, and for long hours inhale the
aroma of their peculiar dreams into a marvelling and bewildered
being.*

—MARCEL PROUST, *The Guermantes Way*

This book has been designed to fill two conspicuous gaps in the available
literature on drugs. First, it has been written as a multi-disciplinary
reference source, presenting botanical, chemical, historical, and neuro-
pharmacological data on hallucinogenic plants. In this respect, this book
has been written for students of the above disciplines and it is hoped that it
will serve to provide such students with easy reference to disparate pieces of
literature, while giving them a broad perspective of the role of hallucinogens in human affairs.

Principally, however, this book has been designed to give access to
reliable, accurate information on hallucinogenic plants to those persons
who use hallucinogens as recreational drugs. Thus far, there has been an
unfortunate reticence on the part of drug plant researchers to provide such
information. As William Burroughs has so aptly put it, "Our national drug
is alcohol. We tend to regard the use of any other drug with special horror.
Anyone given over to these alien vices deserves the complete ruin of his
mind and body" (26).

Alcohol, or ethanol, like many of the alkaloids found in the plants
discussed in this book, is a potent consciousness-altering drug. Alteration

of consciousness through the use of alcohol is not without its attendant dangers—regular, long-term use may result in irreversible brain and liver damage. Addiction to alcohol is commonplace, and very difficult to treat—perhaps 5% of the population of the United States is addicted to alcohol. Further, alcohol may cause pronounced anti-social behavior. A large number of the murders occurring in this country are committed by persons under the influence of alcohol. Excess consumption of alcohol is implicated in most of the highway accidents which claim so many lives each year.

It is not, however, my intention to condemn the use of alcohol. Different people prefer different mind-altering substances. I am no stranger to the chemistry laboratory; I consider alcohol to be a useful solvent, and use it regularly for that purpose, but would no more use it to alter my state of consciousness than I would use butanol or any other solvent. Furthermore, having tried alcohol as a recreational drug, I see no great advantage to be derived from altering my level of consciousness by diminishing it; let alone by running great risks to my health in the process. On the other hand, I have spoken to many people who prefer the effects of alcohol to the effects of one or other of the hallucinogens.

Tobacco, which was one of the most important hallucinogenic plants in use in Pre-Columbian North America (61, 153, 185), is legal in the United States and widely used. It is my opinion that tobacco use in this country today is a classic example of abuse of an hallucinogenic drug. The cultivated strains of the tobacco of commerce are weak, and are used with such regularity as to be devoid of hallucinogenic activity (generally, tolerance to hallucinogenic drugs is quickly acquired). Indeed, it has been my experience that repeated use of low potency marijuana will not result in any appreciable alteration in consciousness either. It seems, therefore, to me, that modern tobacco smokers are exposing themselves to many of the potential dangers of the use of an hallucinogenic substance, owing to the vehicle of administration (i.e. smoking—exposing the smoker to pulmonary carcinomas, emphysema, etc.), without realizing any of the potential benefits to be derived from the use of this substance (such as pleasant alterations of consciousness); all of the nonsense about the enjoyable taste of tobacco notwithstanding.

Again, it is not my intention to condemn the use of tobacco. I am simply trying to put the use, in our culture, of hallucinogenic drugs in the proper perspective. These drugs have been in use as long as alcohol, probably much longer. They are *not* addictive, and no permanent dangers to health are known to result from their use, even over long periods of time (194). It has been pointed out above that habitual use of tobacco may result in lung cancer in some cases—this is not, however, anomalous. Were tobacco used as an hallucinogen (in other words, if it were used infrequently—for only if so used could the plant effectively produce consciousness alteration), this danger would not be present. Clearly, habitual and excessive smoking of anything is likely to expose the smoker to pulmonary problems sooner or later. This example has something to say about routes of administration —obviously, smoking is one of the least healthful ways of ingesting an hallucinogenic plant.

It has been suggested that hallucinogenic drugs produce chromosome damage, short-term loss of memory, impaired resistance to infection, and an "amotivational syndrome." There is absolutely no evidence that hallucinogenic drugs, as they are commonly used, produce chromosome damage (they are certainly *not* teratogenic) (50, 176, 194), short-term memory loss often results from the use of hallucinogens, but does not persist after the effects of the drugs have worn off; recent evidence shows that marijuana, the drug alleged to impair resistance to infection, has no inhibitory effect on immune responses (165), and may even confer *increased* immunity to certain types of tumors (118); and it is farcical to suggest that the use of drugs of any kind *causes* amotivational behavior. Clearly, as Andrew Weil has pointed out, drug use can be a *symptom* of lack of motivation—human beings have, through the years, found hundreds of other ingenious activities such as watching television, playing cards, etc., to further express lack of motivation (194).

Even should some of the theoretical dangers which have been ascribed to the use of hallucinogens be proven by scientific research, this use would still likely be far safer than the use of alcohol. Yet alcohol is legal and many of the hallucinogens are not. Experience shows us that it is neither wise nor possible to prohibit specific types of consciousness alteration by legislation. The only prudent course for a government to follow is to attempt to educate its citizenry as to the merits and dangers of the various modes of mind alteration, and to regulate the production and sale of the substances used for this purpose, to ensure safety. This would entail no cost to society; indeed, consciousness-altering substances are an excellent source of revenue, through taxation which people are, in most cases, happy to bear. Senseless legislation, the detection, arrest, prosecution, and punishment of drug "offenders" represent staggering costs to society, and a gross waste of human time. Prohibition of drugs may even increase their use, it alienates the users from authority, it is counterproductive and completely unnecessary.

This book represents a step toward further educating people about drugs. The majority of the plants discussed in this book produce a state of "benign narcosis" (Burroughs) which is often very pleasurable. Used intelligently, they are quite safe, and often produce new perspectives, beneficial insights into life. Clearly, hallucinogens are popular, and will be used, in spite of legislation. The user of hallucinogens is far safer using plant material, *if the plant can be properly identified,* than pills purchased on the black market, which have been represented to be hallucinogenic drugs. Neither the buyer nor (usually) the seller has any idea what these pills actually contain. Prohibition of hallucinogenic drugs has resulted, in recent years, in numerous deaths caused by poisoning resulting from the ingestion of spurious preparations. Just how widespread is deception in black market drug trade is indicated by the following data on samples of alleged psilocybin or *Psilocybe* mushrooms of underworld commerce, which have been analyzed by Pharm Chem Labs in Palo Alto. Of 333 samples of capsules or mushrooms alleged to contain psilocybin, 25% were inert, 53.7% contained LSD, 1% contained PCP, 4.2% contained LSD and PCP, and 1% contained other miscellaneous compounds. Only 15% actually contained psilocybin. Most of these genuine samples were brought to the laboratory by the persons who had actually picked the mushrooms. Those mushrooms sold on the street were typically *Agaricus bisporus* (the mushroom of commerce) adulterated with LSD, PCP, or nothing at all. Bruce Ratcliffe has proposed the binomial *Pseudopsilocybe hofmannii* for such mushrooms (144).

The reticence, on the part of drug researchers, to publish layman-oriented information on hallucinogens has resulted in the theft, from libraries all over the country, of countless papers and books dealing with hallucinogenic plants. Being a researcher in this field, I know only too well the frustration of finding that articles have been cut out of journals, that whole volumes of scientific literature are missing from libraries. Most likely, those persons who have stolen this material do not have the training required to understand it. If general information on these plants were available, I believe this piracy would in a large measure cease, and this material would remain available to scientists like myself who have a use for it. I feel an obligation to furnish this general information. This book, then, is designed to contribute toward a safer, more intelligent use of hallucinogenic drugs.

The reader will notice that toxicity data are not presented for many of the plants. This is because reliable information is not available at this time. The reader who wishes to use an unfamiliar plant would be well advised to exercise extreme caution. If there is some doubt about identification, an expert should be consulted; hopefully it will be possible to find someone who will cooperate.

This book is by no means comprehensive. Focusing only on North America (Canada, United States, Mexico), it deals mostly with the more important plants. In general, only those plants have been included, whose chemistry has been well characterized; the exceptions having been added to lend taxonomic and geographic breadth to the material. "Hallucinogenic," in the context of this book, refers to plants whose principal physiologic activity, and primary ethnologic importance is the induction of hallucinatory states of consciousness.

PART ONE

Hallucinogenic Plants
of North America

*Here was a panacea . . . for all human woes; here was the secret of
happiness about which philosophers had disputed for so many ages,
at once discovered; happiness might now be bought for a penny and
carried in the waistcoat pocket; portable ecstacies might be had
corked up in a pint bottle; and peace of mind could be sent down in
gallons by the mail coach.*

—THOMAS DEQUINCEY,
Confessions of an English Opium Eater

AGARICACEAE

AMANITA MUSCARIA (Fr. ex L.) Hooker
PLATES 1 AND 5

BOTANICAL DESCRIPTION: Pileus 80–240 mm broad, stipe 80–150 mm long; 20–30 mm thick. Color of pileus yellow, orange, or red; usually with white warts. Gills free, crowded, white, with extremely fine hairy edges. Stipe covered with fine silky white hairs, volva prominent and bulbous; several concentric rings on the stipe directly above the volva. Flesh of pileus and stipe white and firm throughout. Spores 8–11 x 6–8 μ, ellipsoid, spore print white (121).

HABITAT AND SEASON: This mushroom is cosmopolitan, found throughout the continent. *A. muscaria* always grows in a mycorrhizal association, usually with birches and conifers; therefore this fungus is found where these trees abound, principally in cool, mountainous zones. Fruits from early spring through late fall in the United States; in Mexico fruits from summer to fall, in the pine-oak forest belt (121, 180).

HALLUCINOGENIC AGENTS: Muscarine has been isolated from this species; although parasympathomimetic, it is present in trace quantities, insufficient to account for the toxicity of these mushrooms (63). Two known psychoactive principles are ibotenic acid and muscimol, its decarboxylation product (63, 64, 65, 184). Muscazone is present in trace quantities (63). Early reports of the presence of bufotenine and atropine have been shown to have been in error (19, 28, 149, 174, 179).

HISTORY OF ETHNIC USE: R. Gordon Wasson has hypothesized that *A. muscaria* was the euphoriant plant "Soma" of the ancient Aryan invaders of India. Graphic descriptions of the use of Soma are presented in the *Rg Veda*. If this theory is correct, this mushroom has been in use for around 4000 years (189). John Allegro has proposed, on the strength of linguistic evidence, that modern Christianity originated in a cult based on the ritual ingestion of *A. muscaria* (5). While Wasson's theory has been well received, that of Allegro has been severely criticized by other etymologists. The use of this species as a shamanic and recreational inebriant has been observed among Siberian tribes in modern times. The users were seen to ingest several mushrooms, and their urine was subsequently drunk by other celebrants, who also became intoxicated (189). Evidently, the psychoactive principle of *A. muscaria* passes into the urine of the user (131). This mushroom is used today in northern California, especially in Marin and Sonoma counties, particularly by persons who do not know how to identify the more popular *Psilocybe* species. *A. muscaria* is also used recreationally in western Washington, and some users prefer this species to the various psilocybin containing mushrooms. Users seem to prefer the effects of a closely related mushroom, *A. pantherina* (130, 195). Significantly, studies indicate that *A. pantherina* contains considerably higher quantities of ibotenic acid and muscimol than *A. muscaria* (13, 32, 130). Ibotenic acid has also been detected in *A. cothurnata* and hybrids of *A. gemmata* and *A. pantherina* (13, 32, 33). There are no reports of intentional use of either of these species, but poisonings have been reported (43).*

*A. muscaria is called "beni-tengu-take" in Japan (scarlet long-nosed goblin mushroom), and its intoxicating properties are said to be well known in some areas. In parts of Japan, large quantities of *A. muscaria* are dried, pickled, then repeatedly washed prior to ingestion as a food source. This treatment renders the mushroom non-toxic, as the toxins are water soluble. For details, see: R. Imazeki, "Japanese Mushroom Names," *The Transactions of the Asiatic Society of Japan,* Third Series, Vol. XI, 1973.

A similar culinary use of *A. pantherina* has been reported from Washington. In this case, the mushrooms are parboiled, and the water is discarded. The mushrooms are then canned for future use. It is interesting to note that in Washington, the users are unaware of the toxicity of fresh specimens (W. S. Chilton, personal communication).

AMANITA PANTHERINA (D.C. ex Fr.) Krombh.
PLATE 4

BOTANICAL DESCRIPTION: Pileus 50–120 mm broad, stipe 60–120 mm long; 10–25 mm thick. Color of pileus tan to dark chocolate brown with soft white warts, margin striate. Gills white, free, close, with scalloped edges. Stipe white with firm white flesh throughout. Annulus persistent, membranous. Volva non-saccate, firm and bulbous, with a collar where stipe and volva meet. Spores 9–12 × 6.5–8 μ, elliptical, spore print white (121).

HABITAT AND SEASON: This is a cosmopolitan species, widely distributed in North America. Like all species of the genus, *A. pantherina* grows in a mycorrhizal association, usually with conifers and hardwoods (121). It is often found in great abundance in the Pacific Northwest, where it fruits in the spring.

HALLUCINOGENIC AGENTS: Like *A. muscaria,* the principal psychoactive agents of this species are ibotenic acid and muscimol, its decarboxylation product. North American specimens of *A. pantherina* appear to contain the highest concentrations of these alkaloids (13, 33).*

HISTORY OF ETHNIC USE: Despite its chemical similarity to *A. muscaria,* there is no evidence that *pantherina* has ever been used in a magico-religious context. Mycologists have long considered this to be a deadly poisonous species, although there has been only one death attributed to its ingestion, that of an aged man with heart trouble. *A. pantherina* is the most common cause of accidental mushroom poisoning in the Pacific Northwest (143). Such poisonings frequently result in hospitalization and, often as not, the unfortunate victims are treated with atropine sulfate, in the mistaken belief that muscarine is the toxic agent responsible for the effects (132). Atropine, however, potentiates the effects of ibotenic acid and muscimol, and will invariably intensify the experience of *A. pantherina* intoxication. In recent years, owing to popular awareness of the psychotropic potential of *A. muscaria,* this chemical and taxonomic relative has come into widespread recreational use in Washington, Oregon, and California (132, 195). Users generally prefer *pantherina* to *muscaria,* as it is more potent. While some psychedelic mycophagists prefer the psychoactive *Amanita* species to psilocybian mushrooms, most consider them to be inferior, and not a few dislike them intensely (130). *Caution should be exercised with dosage, as this mushroom can be extremely potent.*

* The chemistry of this species is complex. Trace amounts of muscarine have recently been detected in European material (Stadelmann, R. J. et al, *Helvetica Chimica Acta* 59(7): 2432-2436, 1976). Stizolobic and stizolobinic acids have been isolated from Washington collections, and are of unknown pharmacologic activity (Chilton, W. S. et al, *Phytochemistry* 13: 1179–1181, 1974). Scott Chilton and I have detected a yet-unidentified alkaloid in ethanolic extracts of Washington State *A. pantherina.*

PANAEOLUS FOENISECII (Pers. ex Fr.) Kühner

BOTANICAL DESCRIPTION: Pileus 10–25 mm broad, stipe 40–100 mm long; 1.5–3 mm thick. Color of pileus dark smoky brown to reddish brown, surface dry and smooth, margin smooth, may become somewhat striated with age. Gills adnate, dark purple brown to brown, close, broad, often mottled, with whitish edges. Stipe white to pink-brown, brittle, slightly enlarged at base, with tiny hairs at apex. Flesh light brown, thin and watery throughout. Spores 11–18 x 6–9 μ, elliptical, spore print dark purple brown (121).

HABITAT AND SEASON: Grows abundantly on lawns and grassy areas. Must be found in the early morning, as this mushroom wilts and fades away before the midday sun. Cosmopolitan, occurring throughout the continent from spring to fall (121).

HALLUCINOGENIC AGENTS: The presence of psilocybin/psilocin in specimens from Ontario and Indiana has been demonstrated, while specimens from the Seattle area were devoid of hallucinogen content (145). I have ingested a large number of specimens from the Olympia area, with no noticeable effect (130).*

HISTORY OF ETHNIC USE: There is no evidence to indicate that this mushroom was used ritually anywhere in North America. Its use has not been documented in modern times. Ola'h classes this species as "psilocybin latent," that is, specimens may or may not contain psilocybin (129). Chemical studies seem to show very low levels of psilocybin/psilocin in this species (145).

* Recent analysis of specimens of *P. foenisecii* from the federal district of Mexico failed to detect the presence of psilocybin (J. Ott and G. Guzmán, *Lloydia* 39: 258–260, 1976).

PLATE 1. *Amanita muscaria*

PLATE 2. *Psilocybe pelliculosa*

Deadly Poisonous Mushroom

PLATE 3. *Amanita phalloides* (Fr.) Secretan

Cap viscid, pale yellowish-green to greenish, no warts. Volva is persistent and saclike, never flattened to the stipe, as in *A. muscaria* and *A. pantherina*. Fruits in Spring and Fall, usually mycorrhizal with European trees (121).

Opposite, above
PLATE 4. *Amanita pantherina*

Opposite, below
PLATE 5. *Amanita muscaria*

PLATE 4

PLATE 5

PLATE 6. *Lophophora williamsii*

PLATE 7. *Ariocarpus fissuratus*

PLATE 8. *Lophophora williamsii* and mescaline crystals

PLATE 9. *Panaeolus subbalteatus*

PLATE 10. *Panaeolus subbalteatus*

PLATE 11. *Peganum harmala*

PLATE 12. *Argyreia nervosa*

Deadly Poisonous Mushroom
PLATE 13. *Galerina autumnalis* (Pk.) Smith and Singer
Gills attached, rusty brown, spores rusty brown. Annulus persistent, thin, white and hairy. Cap viscid, dark brown when moist, margin striated. Flesh thick, light brown, watery. Grows scattered or abundant on well decayed hardwood or conifer logs (may, however, be on buried wood). Fruits in late Fall, sometimes in early Spring (121).

Opposite above
PLATE 14. *Psilocybe stuntzii*

Opposite, below
PLATE 15. *Psilocybe stuntzii*

PLATE 14

PLATE 15

PLATE 16. *Cannabis sativa.*

PLATE 17. *Psilocybe cyanescens*

PLATE 18. *Psilocybe baeocystis*

PLATE 19. *Psilocybe zapotecorum*

PLATE 20. *Psilocybe cubensis*

PLATE 21. *Psilocybe cubensis*

PLATE 22. *Psilocybe cubensis*

PLATE 23. *Psilocybe cubensis*

PLATE 25. *Psilocybe semilanceata*

Opposite
PLATE 24. *Psilocybe semilanceata*

PLATE 26. *Psilocybe semilanceata*

PLATE 27. *Psilocybe caerulescens*

PLATE 28. *Psilocybe caerulescens*

PANAEOLUS SPHINCTRINUS (Fr.) Quélet

BOTANICAL DESCRIPTION: Pileus 15–30 mm broad, stipe 40–80 (100) mm long; 2–3 mm thick. Color of pileus greenish grey to brownish grey, often with a dark brown disk at the apex, or a blackish ring close to the edge. Gills adnate, broad, at first grey, then blackish and mottled with white edges. Stipe hollow, greyish to reddish brown, somewhat enlarged at the base. Flesh of pileus thin, same color as the surface, odorless. Spores 13–19 x 9–12 μ, lemon shaped, spore print black (129, 159).

HABITAT AND SEASON: These mushrooms are found throughout temperate parts of Canada, the United States, and Mexico, fruiting from spring through fall. This species is particularly common in the mountainous regions of Mexico, in the states of Morelos, Oaxaca, Puebla, Mexico, and Veracruz. Almost always found in cow dung, in groups of from 5 to 15 individuals, in pastures and fields where cattle graze (129, 159, 180).

HALLUCINOGENIC AGENTS: These mushrooms are known to contain the substituted indole alkylamines psilocybin/psilocin, found in many species of the genus *Panaeolus*. Ola'h lists this species as "psilocybin latent," that is, some specimens have tested psilocybin positive, others negative (129). I have tested several specimens of the species from Washington, and they have been found to be devoid of hallucinogenic activity (130).

HISTORY OF ETHNIC USE: There is some controversy regarding this particular species—it may be used ritually in Oaxaca, Mexico (159). This species was one of the first positively identified hallucinogenic mushrooms to be collected in Mexico (150). It later proved, however, to be a species of minor importance, if in fact it is used at all, which some researchers doubt.* There is no evidence to indicate that this species has been used as a recreational drug in the United States in modern times, although it seems that many people have tried it—perhaps those mushrooms found in the United States are inactive as hallucinogens. There is some degree of taxonomic confusion between this species and *P. campanulatus*.

* Dr. Gastón Guzmán (personal communication), despite nearly two decades of study, has seen no evidence of ritual use of *P. sphinctrinus* anywhere in Mexico. Recent analysis of specimens from the Mexican state of Puebla failed to detect the presence of psilocybin (J. Ott and G. Guzmán, *Lloydia* 39: 258–260, 1976). This species is used for recreation in Oregon, and analysis by Pharm Chem has shown "traces of psilocybin." Users will take up to 250 carpophores to produce a strong hallucinogenic effect. (M. C. Dunham, A. T. Weil, and G. Guzmán, personal communications).

PANAEOLUS SUBBALTEATUS
(Berk. and Br.) Sacc.
PLATES 9 AND 10

BOTANICAL DESCRIPTION: Pileus 20–50 mm broad, stipe 50–80 mm long; 2–3 mm thick. Color of pileus dull fawn to dull reddish brown or dark brown, black in maturity, margin marked by a narrow, dark zone. Gills adnate, brownish red to black, close, broad, with grey to white serrated edges. Stipe brittle, stringy, hollow, surface grooved, especially at the top, covered with fine white powder, beneath which the flesh is the same color as the pileus. Flesh of pileus white to yellowish or brownish. Sometimes stains blue where bruised. Odor and taste like commercial mushrooms. Spores 36–45 x 6–7 μ, spore print blackish, often with a purple hue (129).

HABITAT AND SEASON: Grows in clumps and sometimes rings on open ground where composted, on dung, on compost, lawns, straw piles. Fruits from spring to fall. In Mexico, grows from 1300 to 2200 m, principally in the temperate zone. Common throughout the continent (73, 129).

HALLUCINOGENIC AGENTS: These mushrooms contain psilocybin/psilocin, indole alkylamines relatively common to mushrooms of the genus *Panaeolus* (129).

HISTORY OF ETHNIC USE: There is no indication that mushrooms of this species were used in ancient times, in any part of the continent, for ritual or medicine. *Panaeolus subbalteatus* is used as a recreational drug in western Washington, and is even cultivated for this purpose. It has been suggested that this species is also used as a recreational drug in New England (139).

PSILOCYBE BAEOCYSTIS Singer and Smith
PLATE 18

BOTANICAL DESCRIPTION: Pileus 14–54 mm broad, stipe 50–70 mm long; 2–3 mm thick. Color of pileus olive brown to brown, viscid pellicle, staining greenish-blue where injured. Gills dark cinnamon to purplish with whitish edges, adnate, close. Stipe white, stuffed with loose fibrils, staining greenish-blue where injured. Spores 11–12 × 6.3–7 μ, spore print greyish purple (164).

HABITAT AND SEASON: Grows solitary or in clumps on decaying peat moss, wood chips, occasionally on lawns. Fruits from fall through early summer, from northern California through British Columbia (130, 164).

HALLUCINOGENIC AGENTS: This mushroom contains the hallucinogenic indole alkylamines psilocybin and psilocin. Baeocystin and norbaeocystin, analogs of psilocybin, were first isolated from this species, but are of unknown pharmacological activity. *

HISTORY OF ETHNIC USE: There is no evidence of traditional use of this mushroom among Pacific Northwest indigenous groups. In recent years, *P. baeocystis* has become popular among recreational users of hallucinogenic mushrooms in California, Oregon, Washington and British Columbia, who esteem this species as one of the most potent of the psilocybian fungi (130).

---

* Although first isolated in 1968 by Leung and Paul (*J. Pharm. Sci.* 57: 1667–71, 1968), as yet the pharmacology of these interesting psilocybin analogs has not been explored. It is probable that both baeocystin and norbaeocystin are hallucinogenic, and may contribute to the psychoactivity of psilocybian fungi. In addition to *Psilocybe baeocystis*, baeocystin has been found in *P. cyanescens, P. cubensis, P. pelliculosa, P. semilanceata, P. silvatica, P. stuntzii, Conocybe smithii, C. cyanopus* and *Panaeolus subbalteatus*, in concentrations as high as 0.17% of dry weight (Repke, D. B., et al, *Lloydia* 40: 566–578, 1977).

PSILOCYBE CAERULESCENS Murrill
PLATES 27 AND 28

BOTANICAL DESCRIPTION: Pileus 20–88 mm broad, stipe 40–122 mm long; 2–10 mm thick. At first dark brown to dark olive, the pileus fades to a paler hue from the center outward, often becoming chestnut color when moist. Pileus is viscid, often with a striated margin. Gills are adnate, greyish with a slight white fringe. Stipe is hollow, with a soft glassy white covering which breaks up giving a mottled appearance. Flesh of pileus is white to tawny, may be dark brown in upper portion, fleshy, staining blue where injured. Strong odor of grain, astringent acrid taste. Spores 6–8 × 5–6 × 3.5–5 μ, spore print dark purple (164).

HABITAT AND SEASON: Grows in clusters or solitary, in rich earth which has been tilled, in bagasse, and on landslides. Fruits in the summer, during the rainy season. This mushroom was first collected in Alabama in 1923, but has never again been collected in the United States. It seems to be very common in the Mexican state of Oaxaca, from 1200–2100 m altitude (72, 79, 164).

HALLUCINOGENIC AGENTS: This mushroom contains psilocybin/psilocin, hallucinogenic indole alkylamines common to all hallucinogenic species of the genus (79).

HISTORY OF ETHNIC USE: This mushroom is used today by Mazatec curers in Huautla de Jiménez, Oaxaca and environs. The Mazatecs call this species "di-shi-tjo-ki-sho" (sacred mushroom of landslides). *P. caerulescens* is also used by Zapotec curers, near San Agustín Loxicha, Oaxaca, who call this species "razon mbei" (mushroom of reason). Chatino Indians of Oaxaca call this mushroom "cui-ya-jo-o-su" (sacred mushroom of great power) (72, 79). Three varieties of *P. caerulescens* have been described: *mazatecorum, nigripes,* and *albida* (79, 81). This seems to be the most widely distributed hallucinogenic species in the tropical mountains of Mexico, is probably the most widely used, and is surely the species most widely sold to tourists, under the name "derrumbes" (landslides). I have observed Zapotec Indians selling carpophores of this species cured in honey (130). This is not an indigenous practice, although the sacred mushrooms were apparently taken with honey as a flavoring in pre-conquest times (79). The material I examined was a disgusting, fermenting mess, crawling with bugs, and the Indians sold them thus because of the demand of tourists for mushrooms so cured! There is a widespread belief in the United States that the potency of fresh hallucinogenic mushrooms is best preserved by embalming them in honey. I know of no evidence to support this theory, and would wager that the high from the mushrooms I saw would be due rather to the ethanol content than psilocybin (130).

PSILOCYBE CUBENSIS (Earle) Singer
PLATES 20–23

BOTANICAL DESCRIPTION: Pileus 16–80 mm broad, stipe 40–100 mm long; 4–14 mm thick. Color of pileus white to brown, often with pronounced yellowing toward the center, giving a "fried egg" appearance. Gills grey to deep violet or black, adnate to adnexed, close, narrow, with white edges. Persistent annulus, fragile and smooth, white to blue. Flesh white throughout, dry, fibrous, staining blue where bruised or injured. Cap often sticky, no odor. Spores 11.5–17.3 x 8–11.5 x 7–9 μ, ellipsoid, spore print dark yellow brown to dark purple brown.

Synonym: *Stropharia cubensis* Earle (164).

HABITAT AND SEASON: Usually grows in cultivated areas or pastures, most often on cow dung, rarely on horse dung. Also fruits on rich pasture soil, mulch, or straw which has been well mixed with dung. Reported only along the Gulf Coast in the United States, particularly abundant in wet areas of Florida, Texas, and Louisiana, where it is said to be associated with Brahma cattle. Also found in Oaxaca and environs in Mexico. Fruits from February to November or December (79, 164).

HALLUCINOGENIC AGENTS: These mushrooms contain psilocybin/psilocin, substituted indole alkylamines common to mushrooms of the genus *Psilocybe* (12, 79).

HISTORY OF ETHNIC USE: Mushrooms of this species are in use today, by the Mazatec Indians near Huautla de Jiménez, Oaxaca. This mushroom is called "di-shi-tjo-le-rra-ja" (sacred mushroom of cow dung) by these Indians (72). It is evident that at present these sacred mushrooms are used for recreation as well as ritual in Mexico. The Mazatec curers consume from 4 to 8 carpophores to produce an intoxication characterized by auditory and visual hallucinations (72, 79).* Under the name "San Isidro" (the patron saint of the fields), mushrooms of this species, as well as varieties of *P. caerulescens* (which are called "derrumbes," meaning "landslides"), are commonly sold in the Mazatec area as recreational drugs. Some of the natives of this area use these mushrooms as recreational drugs, however, a much higher percentage of these natives collect and sell them to the tourists who come to Huautla for the express purpose of buying hallucinogenic mushrooms. The mushrooms are sold for $.40–$4.00 per "viaje" (trip), or for $8.00–$20.00 per kilo, depending on the distance from Huautla and the bargaining ability of the buyer. Mexican youths will journey to Huautla to buy these mushrooms, and will subsequently sell them for a profit in the schools in Mexico City. This is probably the most widely used hallucinogenic mushroom in the United States (130). Its use has become a *cause célèbre* in Florida, where numerous youths can be seen in

---

*As this species is associated with cattle, it is unlikely that it existed in Mexico prior to the Conquest, when these animals were introduced to this area. R. Gordon Wasson (personal communication) has found that, while some Mexican Indian groups use this mushroom in curing rituals, other groups, who may be aware of its hallucinogenic properties, do not. Wasson has suggested that these groups eschew this mushroom because it has no ancient Pre-Columbian tradition of use, and is therefore regarded to be inferior. The author has observed that, while this mushroom is used ritually in the Mazatec zone, it is regarded to be of inferior quality and may only be used when other varieties are unavailable. This discrimination is doubtless a result of cultural factors rather than chemical differences.

the fields after rains searching for these mushrooms, to the chagrin of many elder members of the populace. Some of these users have required medical attention, doubtless because of the consumption of toxic mushrooms other than *Psilocybe*. I have observed highly unsophisticated users in Dade County ingesting any mushrooms which they may find growing in cow pastures, in hopes that some of the mushrooms so ingested may be hallucinogenic. Fortunately, no deaths have been reported (130, 195).

PSILOCYBE CYANESCENS Wakefield
PLATE 17

BOTANICAL DESCRIPTION: Pileus 20–40 mm broad, stipe 60–80 mm long; 2.5–5 mm thick. Cap color brownish, viscid, fading to yellowish, staining blue where bruised or injured. When moist, cap is chestnut with a striated margin. Gills adnate to decurrent, cinnamon colored, with pale edges. Stipe somewhat enlarged and curved at the base, rigid, whitish and silky. Stipe stains blue when bruised or broken, and may turn completely blue on drying. Spores 9–12 × 5.5–8.3 × 5.7–7 μ, spore print purple brown (164).

HABITAT AND SEASON: Grows on the earth, usually among dead leaves and twigs, especially on decomposing wood chips. First collected in England, this mushroom seems to be common in the state of Washington, at sea level in the Puget Sound region. Fruits from late summer through early winter, often in great abundance (130, 164).

HALLUCINOGENIC AGENTS: Chromatographic studies have demonstrated the presence of psilocybin/psilocin in material collected in Seattle (12).*

HISTORY OF ETHNIC USE: There is no evidence indicating that *P. cyanescens* was used as a ritual or recreational drug by Native Americans. During the last several years this species has come into widespread use in Seattle and Olympia, Washington, and environs. It commonly fruits on "beauty bark"—a bark mulch widely used by landscape gardeners around shrubs and trees—and is frequently found on college campuses, grounds of state office buildings, and in parks, where such landscaping is widespread. It would appear that the electric power utility in Washington has been inadvertently cultivating this species. It has been collected in abundance in several areas on piles of wood chips left by the power company, the detritus of tree limbs which had been impinging on the power lines (130).

* Recent gas chromatographic studies have shown that this is one of the most potent psilocybian species (D. B. Repke and D. T. Leslie, personal communication).

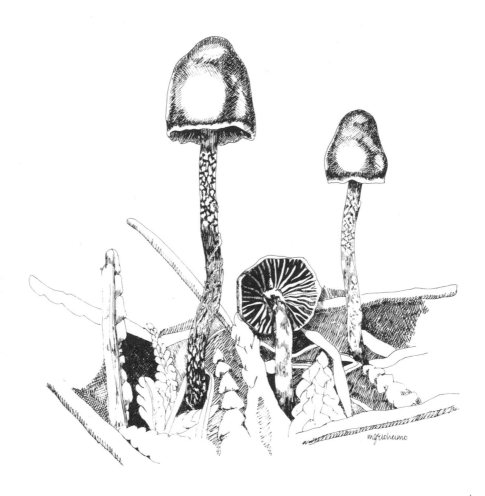

PSILOCYBE PELLICULOSA
(Smith) Singer and Smith
PLATE 2

BOTANICAL DESCRIPTION: Pileus 8–15 mm broad, stipe 60–80 mm long; 1.5–2 mm thick at apex. Pileus smooth and viscid, colored a dark yellow brown which fades to a pinkish buff with green-grey tints with age. Translucent at margin with striations when moist. Gills dull red brown, adnate, narrow, close, with pallid edges. Stipe enlarged at base, pale white to grey, brown flesh covered with a whitish membrane varying in thickness. Flesh of pileus thin and rubbery, flesh of stipe tough, both staining blue or green when injured. Vague musty odor. Spores 9.3–11 x 5–7 μ, spore print dark purple (164).

HABITAT AND SEASON: Found on debris and on humus in and around conifer forests. Scattered or in clumps. Fruits from September to December. Specimens have been found in California, British Columbia, Idaho, Washington, and Oregon (130, 164).

HALLUCINOGENIC AGENTS: Chemical studies have shown that these mushrooms contain psilocybin/psilocin, hallucinogenic substituted indole alkylamines common to the genus *Psilocybe* (178).

HISTORY OF ETHNIC USE: There is no indication that these mushrooms were used in ancient times by the Native Americans of California or the Pacific Northwest. This species was not known to be hallucinogenic until the work of Heim and Wasson, Singer and Smith, and others established that native tribes of Mexico used hallucinogenic mushrooms of the genus *Psilocybe* in curing rituals. Subsequent chemical studies showed that this species is hallucinogenic. *P. pelliculosa* has recently become popular among the young people of western Washington, who use it as a recreational drug. So abundant is this species in western Washington that large collections can often be made. Students in Washington are known to have occasional mushroom parties, and I know of one instance when 150 persons were simultaneously intoxicated by this species. *P. pelliculosa* is often called "Liberty Caps" throughout the Pacific Northwest, and is often confused with *Psilocybe semilanceata* (17, 130).

PSILOCYBE SEMILANCEATA (Fr.) Kummer
PLATES 24–26

BOTANICAL DESCRIPTION: Pileus 5–10 mm broad, stipe 70–110 mm long; 1.5–2 mm thick. Color of pileus pale clay, often tinged with an olive green color. Pileus cone-shaped with a distinct pointed umbo. The margin of the pileus has faint striations, smooth when dry, sticky when moist. Gills crowded, adnate to adnexed, cream to purple brown in color, with whitish edges. Stipe slender, usually wavy, surface smooth and somewhat paler than pileus, whitish near the apex, often greenish blue at the base. Flesh of the pileus white and thin, staining blue-green when injured. Odor of flour. Spores 12–14 x 7–8 μ, spore print purple brown (102, 104).

HABITAT AND SEASON: Fruits in large groups from September to December on lawns, along roadsides, in clearings and pastures, in open forests, and margins of woods. Cosmopolitan, reported in North America from British Columbia, Canada, and Oregon in the United States (102, 104, 130).*

HALLUCINOGENIC AGENTS: Mushrooms of this species contain reportedly high concentrations of psilocybin/psilocin, hallucinogenic indole alkylamines common to the genus *Psilocybe* (91).

HISTORY OF ETHNIC USE: There is no evidence of any ancient ritual use of this particular species in the United States. There are many reports of its use today, under the name "Liberty Caps," as a recreational drug. Students in British Columbia reportedly use this mushroom (62, 130). Andrew Weil describes *P. semilanceata* as the most popular of the hallucinogenic mushrooms in use in western Oregon (195). Despite its small size, the popularity of this species derives from high potency. This species is often confused with *P. pelliculosa* in Washington, British Columbia, and Oregon. Although *pelliculosa* seems to be much more common in these areas, many mushroom users call their "Liberty Cap" collections *semilanceata*.

*The existence of *Psilocybe semilanceata* in the United States is a matter of controversy among mycologists, some of whom maintain that this species occurs only in Europe. Specimens of the mushroom depicted in this book as *P. semilanceata*, collected in Oregon and Washington, have been deposited in the herbarium of Escuela Nacional de Ciencias Biológicas in Mexico City, and have been identified as *P. semilanceata* by Dr. Gastón Guzmán.

PSILOCYBE STUNTZII Guzmán and Ott*
PLATES 14 AND 15

BOTANICAL DESCRIPTION: Pileus 5–35 mm broad, stipe 35–75 mm long; 1.5–5 mm thick. Color of pileus rich chocolate brown, fading to buff when dry; pellicle viscid when wet. Color of stipe is whitish, the white being a powdery covering which rubs off, revealing a brownish-yellow color beneath. Gills widely spaced, narrow, sinuate-adnate to short decurrent, darkening from pale tan in young specimens to chocolate brown in

* The Latin description, chemical assay, and notes on use of this new species have been published in Mycologia (68: 1261–1267, 1976). This mushroom has been misidentified as *P. cyanescens* (A. T. Weil, "Mushroom Hunting in Oregon," *Journal of Psychedelic Drugs*, Vol. 7, No. 1, 1975; and R. and K. Haard, *Poisonous and Hallucinogenic Mushrooms*, Cloudburst Press, 1975) and as *P. pugetensis* (R. Harris, *Growing Wild Mushrooms*, Wingbow Press, 1976).

older specimens. Flesh of pileus waxy, same color as surface or whitish; stipe fibrous and tough, flesh same color as surface. Persistent annulus, which is membranous, usually purple from spore deposits. Stipe turns blue where bruised or handled, pileus often turns greenish-blue at the margin when handled. Odor of grain. Spores 9.3–10.4 x 6–7.1 x 5.5–6.6 μ, spore print dark purple brown (74, 130).

HABITAT AND SEASON: Grows on bark mulch, on lawns, and in fields. Solitary or in dense clusters. Thus far, reported only from western Washington and British Columbia. Fruits from August to December (130).

HALLUCINOGENIC AGENTS: My studies have shown that carpophores of this species contain the hallucinogenic indole alkylamine psilocybin (130).

HISTORY OF ETHNIC USE: There are no ancient reports of the use of this species anywhere in the continent, for medicine or ritual. This mushroom was first collected in Seattle in 1972, where it was fruiting abundantly on the campus of the University of Washington. The mushrooms were usually found on the cedar bark mulch which the gardeners spread around landscaped areas of the campus. Intrepid student experimenters soon learned that these mushrooms were hallucinogenic, and their use as recreational drugs became very popular. These students were admonished, by an article in the student newspaper, that the mushrooms were "dangerous," and the gardeners were instructed to destroy any specimens, and to put fungicides on the mulch! No illness resulting from the use of these mushrooms was reported. In 1974 they fruited abundantly in Olympia and Tumwater, Washington, and again on the campus of the University of Washington. Students in Olympia quickly learned that these mushrooms were hallucinogenic, and they too began to eat them for recreational purposes. Again, no illness resulting from this widespread use was reported. By the fall of 1975, this species was being collected and used in diverse locations in western Washington, and was also collected in Vancouver, British Columbia (130).

PSILOCYBE WASSONII Heim

BOTANICAL DESCRIPTION: Pileus 20–35 mm broad, stipe 30–100 mm long; 2–7 mm thick. Pileus colored brown to rusty red, not viscid, striated at the margin when moist. Gills dark reddish brown, somewhat narrow, sinuate-adnate, with pale edges. Stipe tapers upward, flesh colored to whitish, hollow, usually the same color as the pileus, especially in the lower portion. Flesh of both stipe and pileus turn blue where bruised or injured, smell of raw grains, no distinctive taste. Spores 7–8 x 4.2–5 x 3.9–4.7 μ, spore print brownish purple.

Synonym: *Psilocybe muliercula* Singer and Smith (164).*

HABITAT AND SEASON: Grows solitary or in clusters in ravines or pine-covered mountains at about 3000 m elevation. Fruits in late August or September. Has been found only near Toluca, Mexico (79, 164).

HALLUCINOGENIC AGENTS: This mushroom is known to contain the substituted indole alkylamines psilocybin/psilocin, hallucinogenic agents common to this genus (12, 79).

HISTORY OF ETHNIC USE: This mushroom has been observed in use by Mexican curers, for the purpose of divination and curing (79). Its use is often attended by elaborate rituals. Curiously, Singer and Smith and (separately) Herrera obtained these mushrooms in the marketplace of Tenango, suggesting that perhaps they are used for recreation by some natives of this region (79, 163). Ordinarily, hallucinogenic mushrooms are not an article of commerce—although fungi of many species are sold in the markets, the sacred mushrooms are not generally seen—they are handled under the counter and their use is a serious business. This mushroom may have been used by the ancient Aztecs, as "teonanácatl." Nahuatl Indians today call this mushroom "siwatsitsintli" in Nahuatl, or "mujercitas" or "señoritas" (little women, little girls) in Spanish (79).

* Under the international rules of botanical nomenclature, the name *P. muliercula* takes precedence, as it was the first name published with a Latin description. Heim first described this mushroom in French in the *Comptes rendus* of the Académie des Sciences on 18 November 1957 (245:1761), and announced his intention officially to name the species after the Wassons, who first penetrated the sacred mushroom cult of Mexico. Singer had read some of Heim's preliminary publications on the Mexican mushrooms, and finally made a whirlwind two week trip to Mexico (his first) in 1957, following in the footsteps of the Wassons and Heim, even speaking to their informants. He learned of the existence of this mushroom from one of Heim's collaborators, and bought specimens in the marketplace of Tenango del Valle on 30 July 1957. *He did not collect this mushroom in the field.* He and Alexander Smith hastily published a Latin description of this new species in *Mycologia* (50:141), which appeared on 2 April 1958, less than a month before Heim's Latin description (*Revue de Mycologie* 23:119, appearing 29 April 1958), giving it the name *P. Wassonii.* Singer and Smith subsequently published a lengthy discursion (163, 164) which added nothing to our knowledge of the Mexican sacred mushroom cults, knowledge gained from the diligent work of the Wassons and Heim, to whom they gave no credit. Singer was aware of Heim's intention, and his discourtesy to a senior colleague is unforgivable. Knowing the details of this episode, I cannot in good conscience accept the name *P. muliercula,* as scientific convention would dictate. For me our mushroom is and will remain *P. Wassonii.* After having taken this stance in the first edition of this book, I was excoriated by Alexander Smith in a "brief article" in *Mycologia* (69: 1196–1200, 1977). In my reply to Smith's ill-conceived objections, I defended my position and offered further details. See: *Mr. Jonathan Ott's Rejoinder to Dr. Alexander H. Smith.* Ethnomycological Studies No. 6. Botanical Museum of Harvard University, 1978.

PSILOCYBE YUNGENSIS Singer and Smith

BOTANICAL DESCRIPTION: Pileus 16 mm broad, stipe 55 mm long; 2 mm thick at apex, 3 mm thick at base. Color of pileus deep rusty brown marked with buff grey striations at the margins, slightly viscid. Gills very narrow, crowded, adnate, grey buff with white edges, staining blue where bruised. Stipe dark chestnut brown, becoming auburn at apex, covered with light brownish fibrils. Flesh of pileus and stipe same color as surface, or slightly lighter, staining blue where injured—flesh of stipe often fragile. Spores 5–5.5 x 4.3–5 x 3.6–3.9 μ, spore print dark lilac (164).

HABITAT AND SEASON: Grows in clumps of varying sizes on rotting wood or humus which contains rotting wood. Fruits during summer rainy season. Found in North America only in the Mexican states of Oaxaca, Puebla, and Veracruz (72, 79, 164).

HALLUCINOGENIC AGENTS: This mushroom contains psilocybin/psilocin, indole alkylamines common to the genus *Psilocybe* (12, 79).

HISTORY OF ETHNIC USE: The use of this mushroom by Mazatec Indians in divination and curing ceremonies has been documented (79). Further, it has been postulated that this is the "tree fungus" mentioned by early Spanish chroniclers, allegedly used in an hallucinogenic beverage by Yurimagua Indians in the region of modern Peru. The use was said to be in a ritual context; it has apparently ceased in this area, and has never been observed in modern times (159). The Mazatecs call this mushroom "dinizé-ta-a-ya" or "pajarito del monte," which means "little bird of the mountain" (72).

PSILOCYBE ZAPOTECORUM Heim
PLATE 19

BOTANICAL DESCRIPTION: Pileus 60–110 mm broad, stipe 100–200 mm long; 10–20 mm thick at base. Pileus always twisted and asymmetric, color of pileus from yellowish to purple-brown-black, with dark violet stripes, growing violet black at the margin. Gills close, sinuous, adnexed, violet purple, not very broad. Stipe hollow, hard, fibrous, brown with yellowish interior. Flesh of pileus brown, yellowing in center, turning completely blue when injured. Odor of flour, taste astringent. Spores 6–8.8 x 3.5–5 x 3–4 μ, spore print brown purple (164).

HABITAT AND SEASON: Grows on swampy soil. Fruits from June to September, between altitudes of 900 and 1800 m. Thus far, found only in Oaxaca and Veracruz, Mexico (164, 180).

HALLUCINOGENIC AGENTS: As do all hallucinogenic species of the genus *Psilocybe,* this species contains psilocybin/psilocin (12, 79).

HISTORY OF ETHNIC USE: Chatino Indians of Oaxaca call this mushroom "cui-ya-jo-o-tno" (great sacred mushroom), and use it to this day in rituals. This species is also employed by Zapotec Indians of Oaxaca, who call it "mbey san" (mushroom of the saints) or "piule de barda" (narcotic of Christ's crown of thorns) (79). As these names imply, this mushroom is used in a ritual context, the rituals having been mingled inextricably with Christianity, which was, of course, forced upon the Mexican Indians subsequent to the Spanish conquest. This phenomenon has been extensively documented by Wasson, who dissects a Mazatec mushroom ritual (188). What we are seeing, then, is a curious blend of ancient ritual practice and modern Catholic dogma. No doubt this represents an attempt to perpetuate an ancient custom (the ingestion of hallucinogenic mushrooms) while staying in the good graces of the conquering churchmen. The Indians therefore continue to ingest sacred mushrooms, but attribute the effects to a Christian god in a Christian context. Curers in Mexico who continue to use hallucinogenic mushrooms often have altars in their homes, with representations of Santo Niño de Atocha (a Catholic conception of the young Christ) – this is another example of this phenomenon (130, 188).

ARACEAE

ACORUS CALAMUS Linnaeus

BOTANICAL DESCRIPTION: A stout, erect, perennial herb, with an especially aromatic root. Leaves 40–80 cm long, 0.8–2.0 cm wide, linear and crowded at the base, with a scape which resembles the leaves and bears a lateral spadix 3–8 cm in length and 1 cm in thickness, covered with flowers enclosed in a spath. Flowers dense, yellow-brown and perfect, with a perianth of 6 scale-like sepals. Fruit hard, covered, with a gelatinous interior encapsulating 2–3 seeds (86).

HABITAT AND SEASON: Cosmopolitan, this plant is particularly abundant in swamps and streams of northern and eastern United States and eastern Canada. Grows from May through July (86).

HALLUCINOGENIC AGENTS: Studies in India have led to the isolation of asarone, and its stereoisomer beta-asarone from the roots of this plant. Although psychoactive, there is at present no evidence that these compounds are hallucinogenic. Hallucinosis is apparently induced only by large doses of this plant—perhaps the hallucinogenic agent is a trace constituent which has not yet been identified (11).

HISTORY OF ETHNIC USE: Called "sweet flag," this plant was gathered by Europeans to extract the oil as a perfume base (159). *A. calamus* may have been used as an admixture to the hallucinogenic "flying ointment" used by European witches (34). Roots of this plant were widely used by the Native Americans of Canada and the United States, as a medicine. A decoction of the root was drunk, the fresh root was chewed, or the powdered root was smoked. Apparently, the use of this plant in North America was analogous to the use of coca leaves (*Erythroxylon coca*) in South America—both plants were used to combat fatigue, ward off hunger, and increase stamina. The Cree of Alberta would chew a piece of the root 1–2 inches in length to overcome fatigue; a 10 inch length of root was chewed to produce hallucinosis (88, 159).

CACTACEAE

ARIOCARPUS FISSURATUS (Eng.) Schumann
PLATE 7

BOTANICAL DESCRIPTION: A grey-green cactus, becoming yellowish with age, subglobose, level with, or rising slightly above the ground. May be flat or rounded on top, up to 10 cm in height, 5–10 (up to 13) cm in diameter. Tubercules 10–20 mm long, 15–22 mm broad, basally compressed, with rounded apices, with numerous fissures. Flowers 2.5–4.5 cm in diameter, 1.5–3.5 cm long, outer perianth light magenta, sometimes becoming whitish, inner perianth light magenta. 5–10 stigmas, 1.2–5 mm long, style 1.6–2.1 cm long. Fruit whitish or greenish, 5–15 mm long, 2–6 mm in diameter.

Synonym: *Roseocactus fissuratus* Berger (7).

HABITAT AND SEASON: Grows on limestone hills, along the Rio Grande in southern Texas. Also found in northern Mexico, in the states of Coahuila, Zacatecas, and San Luis Potosi. Flowers from September to October (7).

HALLUCINOGENIC AGENTS: The mescaline analog, N-methyl-3, 4-di-methoxy-β-phenethylamine, has been isolated from the flesh of this cactus, as well as the flesh of *A. retusus,* another species used as an hallucinogen in Mexico. This compound has not been tested pharmacologically. In addition, hordenine and N-methyl tyramine, alkaloids also found in *Lophophora williamsii* and other species of *Ariocarpus,* have also been detected in this plant (116, 126, 128).

HISTORY OF ETHNIC USE: In the United States, this cactus is called "living rock," or "star cactus." The Tarahumara Indians of Mexico use this cactus in ritual, under the name "sunamí." These Indians also use peyote, *L. williamsii,* but consider *A. fissuratus* to be much more potent. This cactus is also used by the Huichol Indians of northern Mexico. Other names for this cactus in Mexico are: "chaute," "chautl," and "peyote cimarrón." This cactus is not very common, and there is no evidence that it has been widely used as a recreational drug in recent times. It has, however, been declared a sacrament by proponents of a modern religious group in California, who cultivate it for use in religious ceremonies (11[4], 130, 159).

CORYPHANTHA MACROMERIS (Eng.) Lemaire

BOTANICAL DESCRIPTION: A small cactus, often polycephalous and basally branched, up to 20 cm in length. Large tubercules of soft density and loose arrangement, 12–30 cm in length, superiorly grooved for about ⅔ of their length. Spines in groups of 10–17, slender, with the radial spines white and several black central spines up to 5 cm in length. Flowers large, up to 8 cm in breadth, purple in color. Fruit 15–25 mm long, with globose seeds which are shaded from brown to yellow, with smooth surfaces (20).

HABITAT AND SEASON: Grows in the vicinity of the Rio Grande in the southwestern United States, and in northern Mexico as far south as Zacatecas. Found in calcareous desert, on rocky hillsides, and in dry river beds (20).

HALLUCINOGENIC AGENTS: The flesh of this cactus contains macromerine, a phenethylamine derivative similar in structure to mescaline, the main active principle of the peyote cactus, *L. williamsii*. Experiments with squirrel monkeys indicated that macromerine is psychoactive (87).

HISTORY OF ETHNIC USE: This cactus is called Doñana in northern Mexico, and may still be used in this area as a ritual hallucinogen. It is apparent that, in recent years, youths in the United States have used this plant as a recreational drug. *C. macromeris* has recently been declared a sacrament of a hallucinogen-oriented church in California, whose members cultivate this plant for use in religious ceremony. At present, no data are available on the toxicity of this plant (115, 130).

LOPHOPHORA WILLIAMSII (Lem.) Coulter
PLATES 6 AND 8

BOTANICAL DESCRIPTION: A small, hemispherical cactus which grows from a thick root. Flesh green, consisting of several (usually 8, but often 5 or 6) sections called ribs which taper upward in a conical fashion, merging at the top of the cactus where they are 5-8 cm broad. These ribs are crowned with dense tufts of short, delicate, compact hairs. When mature, the ribs are broad and rounded, the remnants of the tufts forming a cushion in a small central depression. Flowers are white to rose, small, with 4 stigmas, 15-25 cm broad, located at the summit of the ribs. Ovary naked, free from scales, but downy (39).

HABITAT AND SEASON: This cactus is found in the deserts of northern Mexico and the southwestern United States. Found generally near rivers, particularly the lower Rio Grande in Texas, extending southward, its range includes the Mexican states of Chihuahua, Coahuila, Nueva Leon, Tamaulipas, San Luis Potosi, Zacatecas, and Durango. Has also been reported from New Mexico. Flowers in summer (39).

HALLUCINOGENIC AGENTS: A veritable factory of alkaloids, more than 50 have been detected thus far in the flesh of this plant. Mescaline is without a doubt the principle hallucinogen, isoquinolines such as lophophorine, peyoglutam, and mescalotam doubtless contributing to the complex of symptoms characterizing the intoxication. Contrary to popular belief, this plant *does not* contain strychnine, an alkaloid of an unrelated genus, *Strychnos* (162, 166).

HISTORY OF ETHNIC USE: This cactus was important to the ancient Aztecs as a ritual hallucinogen (159). The Aztecs called this plant "peyotl," meaning "furry thing," a reference to the tufts of hairs crowning the mature plant (45, 46). This cactus is used today by the Tarahumara, Huichol, and Cora Indians of northern Mexico (159). It has been reported that some Yaqui sorcerers use this cactus, and believe that a spirit, called "Mescalito" resides in the plant (27). Although the Spanish conquerors zealously endeavored to eliminate its use in Mexico, it persists to this day—the plant is now called "peyote." The use of peyote has in fact spread to the United States and Canada, where Native American groups utilize the cactus in rituals, incorporated as the Native American Church, to avoid bothersome restrictions. Several of the dried buttons (tops of the cactus—the roots contain only traces of mescaline) are chewed, the intoxication is marked by vivid hallucinations. Extremely bitter tasting, peyote sometimes induces vomiting. Peyote and synthetic and extracted mescaline have been sold on the illicit drug market in the United States during the last twenty years (130). Peyote has received a good deal of attention in scientific and literary circles. Havelock Ellis, S. Weir Mitchell, and Heinrich Klüver made early contributions to the study of this plant. The publication, in 1954, of *The Doors of Perception* by Aldous Huxley stimulated popular interest in this plant.

CANNABACEAE

CANNABIS INDICA Lamarck

BOTANICAL DESCRIPTION: See ensuing description for *C. sativa*. This species is distinguished from *C. sativa* in that the plant is only about 1 m in height, rarely reaching 1.5 m. The leaves are alternate, with darker color and coarser venation than *C. sativa*. The leaflets are much broader and have coarser serrations than *C. sativa*. This plant has a strong odor. *C. indica* is more densely branched than *C. sativa*, has, overall, a conical shape, and a harder stem yielding lower quality fiber than its relative. The seed is 3.5 x 2.4 mm, plump, spherical, with a marbled, mottled coat. At the attachment to the stem, there is a definite abscission layer, and definite articulation at the base of the achene (3, 60).

HABITAT AND SEASON: This species is native to India and Central Asia, and has only recently been introduced to the New World, where it is grown by marijuana users in California, due to its high resin content. It has also reportedly been introduced to Mexico (130, 157).

HALLUCINOGENIC AGENTS: The stem and leaves of this plant contain a complex of compounds called the cannabinols. These compounds are non-alkaloidal, and do not contain nitrogen. The main active principle is Δ^1-3, 4-*trans*-tetrahydrocannabinol (35). Because of a confusion in nomenclature, this compound is often called Δ^9-THC.

HISTORY OF ETHNIC USE: Under the name "bhangas," hemp, probably *C. indica* and *C. sativa*, is first recorded in history in the *Atharva Veda*, written between 1400 and 1000 B.C. It was used as a religious sacrament, along with "Soma," which is also described in the *Atharva Veda* (2, 3). *C. indica* has little history in the New World, and has only recently been accepted by botanists as a species distinct from *C. sativa*. This distinction has led to a number of interesting court cases—many of the marijuana statutes in the United States specify that only *C. sativa* is illegal, and in the dry state, as marijuana is commonly sold, the two species are indistinguishable. Further, botanists have accepted the existence of a third species of hemp, *C. ruderalis*, which is likewise indistinguishable from its relatives in the dry state. The confusion which has resulted has prompted many states to change their laws to specify that any plant of the genus *Cannabis*, or any plant containing the cannabinols is illegal (130).

CANNABIS SATIVA Linnaeus
PLATE 16

BOTANICAL DESCRIPTION: A weedy annual herb which grows up to 5 m in height. Leaflets 6–11 cm long, 0.2–5 cm wide, dark green on upper surface, pale green underneath. Leaves are digitate, opposed at base of plant and in spirals toward the top. Leaflets are sessile, serrated, 8–10 in number, linear to lanceolate and long acuminate. Flowers are green to yellow-brown, about 5 mm long with 5 stamens, and usually appear in pairs. Fruit is an ovoid achene, somewhat compressed and brownish, and bears one ovoid seed which is ash-grey, 3–5 mm long by 2 mm wide (159).

HABITAT AND SEASON: This plant is native to temperate zones of Central Asia. It has been introduced as a cultigen throughout the world. Having escaped cultivation, this plant is now widely distributed (159).

HALLUCINOGENIC AGENTS: The leaves and stem of this plant contain a complex of compounds called the cannabinols. The main active principle is Δ^1-3, 4-*trans*-tetrahydrocannabinol (35). Because of a confusion in nomenclature, this compound is often called Δ^9-THC.

HISTORY OF ETHNIC USE: Considered to be one of the oldest cultigens, "hemp," or "marijuana" has been associated with man in the Old World for thousands of years. It is believed to have been introduced to the New World by the Spaniards in Chile in 1545 (182). The king of England ordered the planting of hemp in Jamestown, Virginia in 1611, the first documented instance of its introduction to the United States (22). Recent evidence has indicated that marijuana may have been introduced to the New World centuries before Columbus, by Phoenecian traders (68). In any case, hemp came to be widely cultivated in the New World for use in medicines, as a source of fiber, and as a source of oil. Hemp extracts were widely used as ingredients in spurious patent medicines sold in the United States around the turn of the century (54). The sale and promotion of such medicines, which often contained large quantities of cocaine and opiates, was in part responsible for the enactment of the Pure Food and Drug Act in 1906, which controlled the distribution and sale of drugs in the United States. Marijuana, however, continued to be legal, and its use came to be associated with blacks (who has brought the tradition of use of hemp from Africa) and Mexican-Americans (who had likewise brought the tradition of marijuana use from Mexico). In a blatantly racist action, designed to discredit and defame blacks and chicanos, who had been providing "unfair" (i. e. cheap) competition to the American labor force in times of economic scarcity, several states began passing anti-marijuana laws in 1914 (3, 83). Marijuana has been illegal throughout the United States since the passage of the Marijuana Tax Act in 1937.

CONVOLVULACEAE

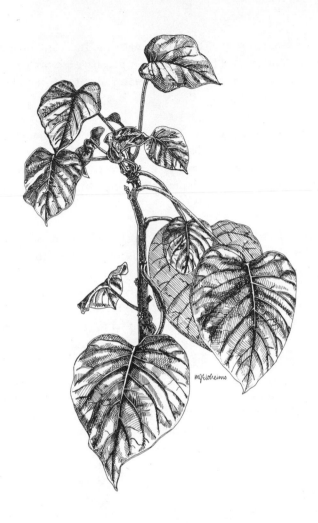

ARGYREIA NERVOSA (Burm.) Bojer
PLATE 12

BOTANICAL DESCRIPTION: A large silvery liana, with white-tomentose stems. Leaves 12 (up to 22) cm long, almost as wide, ovate, cordate, glabrous above, densely white-tomentose below. Flowers rosy purple, corolla 7 cm long, cymose. Corolla funnel-shaped, having 5 petals. 5 stamens, anthers grayish, 5 mm long by 2 mm wide, oblong and straight. Carpels comate, in a four-celled ovary. 5 sepals, greenish-white, 15 mm long by 10 mm wide. Fruit a dry hard berry. Seeds brown, puberulent, 7 mm long by 6 mm wide (117, 140)

HABITAT AND SEASON: This plant is native to India and now exists as a cultigen in some parts of Hawaii, where it may have escaped cultivation. It is sold in plant shops in the United States, due to its popularity as an ornamental (117, 140).

HALLUCINOGENIC AGENTS: The seeds of this plant contain the ergoline compounds common to all hallucinogenic plants of the family *Convolvulaceae,* in higher concentrations than any other species. Lysergic acid amide and other ergoline compounds are present in as large a concentration as 3 mg per gram of seeds (31, 94).

HISTORY OF ETHNIC USE: Called "baby Hawaiian woodrose," the capsules of this plant have long been used, as well as those of *Ipomoea tuberosa* (Hawaiian woodrose) as ornamentals in floral arrangements (they resemble wooden flowers). Recently this use has ceased, owing, apparently, to some legal control, although *I. tuberosa* capsules are still used by some florists. When it was recently discovered that the seeds contained in these capsules were potent hallucinogens, their use became popular on the West Coast and persists to this day—it is still possible to purchase *A. nervosa* plants and cultivate them for their seeds. As little as two grams of seeds produces hallucinogenic effects not unlike LSD, which persist for 8 hours or more. The seeds must be ground prior to ingestion, for if swallowed whole poor digestion and absorption will vitiate the effects (130).

IPOMOEA VIOLACEA Linnaeus

BOTANICAL DESCRIPTION: A highly branched, annual, glabrous vine. Leaves ovate, 4–10 cm long by 3–8 cm wide, acuminate and deeply cordate. Flowers are cymose, 5–7 cm wide, with 3–4 flowers on a cyme. Tube of flower is white, becoming white, blue, red or purple at the limb of the corolla, which is 5–7 cm long. Sepals are 5–6 mm long, acute and subequal. Fruit is black and ovoid, about 13 mm in length.

Synonym: *Ipomoea tricolor* Cavanilles (159).

HABITAT AND SEASON: Native throughout southern and western Mexico; widely planted elsewhere as a garden flower. Grows on hillsides in moist thickets (159).

HALLUCINOGENIC AGENTS: The seeds of this morning glory contain the indole derivative lysergic acid amide and its isomer, as well as other ergoline compounds common to hallucinogenic plants of the family *Convolvulaceae* (89).*

HISTORY OF ETHNIC USE: Called "tliltliltzin" by the ancient Aztecs, the seeds of *I. violacea* were used in medicine and ritual. This use persists in Oaxaca today, the seeds of this plant are now called "badoh negro." Apparently this plant is mainly used by curers when the hallucinogenic mushrooms are not in season (159). Prior to ingestion, the seeds are ground on a metate–if swallowed whole, they are inactive. There are several garden varieties of *I. violacea*–called "Heavenly Blues," "Pearly Gates," "Flying Saucers," and "Wedding Bells," descriptive names curiously suggestive of the potential of these plants to produce altered states of consciousness. When it was discovered that this common garden seed was hallucinogenic, it came to be used as a recreational drug in the United States. About 50 seeds were eaten to produce hallucinosis. Although the seeds are themselves relatively safe, the government of the United States had suppliers poison them with an emetic, in order to discourage their use, foolishly causing users to experience sickness upon ingestion. This is a classic example of the irrational, irresponsible attitude of the American government toward drugs of the hallucinogen class (130).

---

*Chemical work has shown that lysergic acid amide and related alkaloids occur in the leaves and stem of *I. violacea* and *R. corymbosa,* in a dry weight concentration approximately equal to that of the seeds. See: K. Genest and M. R. Sahasrabudhe, *Economic Botany,* 20: 416, 1966. It has been recently found that stems of hallucinogenic morning glories are used along with the seeds in curing rituals in Mexico (Tim Knab, personal communication).

RIVEA CORYMBOSA (Linn.) Hallier

BOTANICAL DESCRIPTION: A large, climbing woody vine. Leaves cordate or ovate-cordate, 5–9 cm long by 2.5–4 cm wide, glabrous to pubescent and long-petiolate. Flowers appear in congested cymes, 2–4 cm long, glabrous, white to whitish with greenish stripes, with 2 stigmas. Sepals scarious and woody, ovate to ovate-lanceolate, around 1 cm long. Ovary glabrous with two cells. Fruit is baccate, indehiscent, ellipsoid, with one seed which is round, woody, puberulent and brownish in color.

Synonym: *Turbina corymbosa* (Linn.) Rafinesque (159).

HABITAT AND SEASON: Native to the Gulf of Mexico coast and tropical North America, appearing in moist or wet thickets and as weeds in hedges (159).

HALLUCINOGENIC AGENTS: The seeds of this morning glory contain lysergic acid amide and other related ergoline compounds.* Ergoline alkaloids are also found in lower fungi of the genus *Claviceps* (ergot), a fact of great chemotaxonomic interest. It was from these compounds that LSD was first synthesized by Albert Hofmann (89).

HISTORY OF ETHNIC USE: This plant was called "coaxihuitl" or "coatlxoxoqui" (snake plant) by the ancient Aztecs (46). The seeds were called "ololiuqui" (round things), and were used as a ritual hallucinogen. This use persists today in Mexico, the seeds of this plant are now called "badoh" (159). In the Maya area, this plant was called "xtabentum" (61). Bas relief carvings of "coaxihuitl" flowers are evident on the statue of Xochipilli, the "Prince of Flowers," probably the god of hallucinogenic flora in the Aztec pantheon, attesting to the ritual importance of *R. corymbosa* in Aztec culture. Also represented in the remarkable carving are caps, in cross section, of "teonanácatl" (hallucinogenic mushrooms), flowers of "sinicuichi" (*Heimia salicifolia*), and flowers of "quauhyetl" (*Nicotiana tabacum*) (186).

*See footnote, page 60.

SOLANACEAE

ATROPA BELLADONNA Linnaeus

BOTANICAL DESCRIPTION: A perennial branched herb which grows up to 1.5 m in height. Leaves 8–20 cm long, entire, acuminate, either alternate or in unequal pairs. Flowers 25–30 mm long, solitary or on rare occasions paired, greenish-purple in color. Corolla has five lobes, is of tubular to bell shape. Five stamens are inserted at base of corolla. Calyx is pentafid, leafy, with triangular-ovate lobes which are acuminate. Fruit is a spherical black berry 15–20 mm in diameter, and contains many seeds (159).

HABITAT AND SEASON: Cosmopolitan, this plant is native to Europe, and was introduced to North America for use in medicine. Has escaped from cultivation, and may now be found in many parts of the continent in shady spots on woody hillsides. Flowers from June to August (159).

HALLUCINOGENIC AGENTS: The roots, leaves, berries and stem of this plant contain the tropane alkaloids scopolamine and atropine/hyoscyamine (stereoisomers) (169). Atropine is commonly used in medicine to counteract the effects of mushroom poisoning, particularly muscarinic poisonings. This is often unfortunate, for atropine potentiates the effects of the toxins of *Amanita muscaria, A. pantherina,* and *A. cothurnata* (143).

HISTORY OF ETHNIC USE: Called "deadly nightshade," or "belladonna" (beautiful woman), extracts of this plant were used in Italy to dilate the pupils of women's eyes, to enhance beauty (61). This plant was used in an ointment by European witches to achieve hallucinosis. It was administered by smearing the ointment on the body, or on a broom handle which was then used for masturbation, the material being thereby applied to the rectal and vaginal mucosa. The combination of this vehicle of administration, with the sensation of flight induced by the tropane alkaloids, produced the traditional Halloween image of a witch flying on a broomstick (77). *Potential users of this plant would be well advised to exercise extreme caution—overdose of parts of this plant may cause sickness, even death.**

*See footnote, page xiii.

DATURA DISCOLOR Bernhardi

BOTANICAL DESCRIPTION: A branched herb with a stem up to 1.5 m tall. Leaves ovate, large, dentate, or almost entire. Stem and leaves almost pubescent. Corolla nearly white, 14–16 cm long, trumpet-shaped, with 10 teeth. Calyx 4–6 cm long, drooping with the corolla, leaving a frill at the base of the capsule. Capsule ovoid, 4–6 cm wide, 6–7 cm long, covered with long slender spines. Seed black (8).

HABITAT AND SEASON: Grows in southern California and New Mexico in the United States, and in northern Mexico and the West Indies. Flowers from April to October (8).

HALLUCINOGENIC AGENTS: In common with all plants of the genus *Datura*, the leaves, stems, roots, and seeds of this species contain the tropane alkaloids scopolamine and atropine/hyoscyamine (159). Scopolamine was used in Germany during the second world war in the interrogation of prisoners, and was considered to be a "truth serum"–this effect was doubtless produced by using large doses and frightening or torturing the unfortunate victim.

HISTORY OF ETHNIC USE: This plant has a long history of use in Mexico and the southwestern United States, as a ritual hallucinogen (159). *Datura* species were used by the ancient Aztecs, who called these plants "toloatzin," or "tolohuaxihuitl" (meaning "drooping head"–a reference to the drooping seed capsules). These were generic names for plants of the genus *Datura*. It is apparent that, at least with respect to medicinal plants, the Aztecs had made efforts toward a botanical classification of the flora of Mexico (45). *Datura* species were also used in the Old World, particularly *D. metel*. This plant was used as an aphrodisiac in the East Indies, as an admixture to *Cannabis* spp. in India, and as an admixture to *Nicotiana* or *Cannabis* spp. in Asia and Africa (159). Recently it has become apparent that, due to the information about the use of *Datura* species in Castaneda's books, many Americans have experimented with these plants as recreational drugs. There have been reports of sickness and hospitalization resulting from this use (130). *Prospective users should use extreme caution, as overdose of this plant may produce sickness or death.**

*See footnote, page xiii.

DATURA INNOXIA Miller

BOTANICAL DESCRIPTION: A branched herb with a stem up to 2 m tall. Leaves and stem covered with soft hairs. Leaves ovate, entire. Corolla white, 15–18 cm long, with 10 teeth, limb 10–12 cm wide. Calyx with uneven lobes. Anthers white, approximately 2 cm long, style 13–17 cm long. Capsule ovoid, nodding, 6–6.5 cm in diameter, covered with long slender spines. Seeds brown.

Synonym: *Datura meteloides* De Candolle ex Dunal (8).

HABITAT AND SEASON: Cosmopolitan, grows from the southwestern United States through Mexico, and in South America. Often cultivated in gardens, as an ornamental. May be found in the eastern and southeastern United States, where it has escaped cultivation (8).

HALLUCINOGENIC AGENTS: The leaves, roots, stems, and seeds of this plant contain the tropane alkaloids atropine/hyoscyamine, and scopolamine, an epoxy derivative. These compounds are often called the "belladonna alkaloids"–they were first isolated from *A. belladonna* (159).

HISTORY OF ETHNIC USE: The Zuni Indians called this plant "a-neg-la-kya." The rain priests would chew the root, and ask the spirits of the dead to persuade the gods to bring rain (159). The ancient Aztecs called this plant "toloatzin," or "tolohuaxihuitl," being variants of the same name which was a generic name for plants of the genus *Datura* (46). This was one of the most popular species in use, in divination and curing. The Tarahumara Indians today add this plant to a beverage called "tesquino" (prepared from sprouted maize) as an aid in the diagnosis of illness by curers (159). Castaneda has reported that certain Yaqui sorcerers use this plant in an ointment which, applied to the genitals, legs, and feet, induces a sensation of flight. These sorcerers call this plant "toloache," a name derived from the Aztec language, but more commonly use the epithet "hierba del Diablo" (Devil's weed), a name derived from the Spaniards (27). *Datura* species were called by this name in Europe, due to their use in the witches' ointment which was likewise said to produce a sensation of flight (77). The European witches were also known to smear this ointment on the genitals and lower extremities. It is tempting, therefore, to assume, despite the long history of the use of hallucinogenic plants of the genus *Datura* in the New World, that this reported use by the Yaquis is not indigenous, and was derived from European influence in recent times (130). Indeed, the Yaqui sorcerers also use *Cytisus (Genista) canariensis* as a ritual hallucinogen–this is an Old World plant introduced by the Spaniards to Mexico (159). Reports of the use of this plant by the Yaquis have led to the belief that the flowers of *C. scoparius* (Scotch Broom), a common plant growing abundantly in cleared areas of the Pacific Northwest, are hallucinogenic–this belief seems unfounded, however, this plant has not been investigated pharmacologically (130). *Prospective users of* Datura *plants should be very cautious–overdose of these plants can cause sickness, even death.**

*See footnote, page xiii.

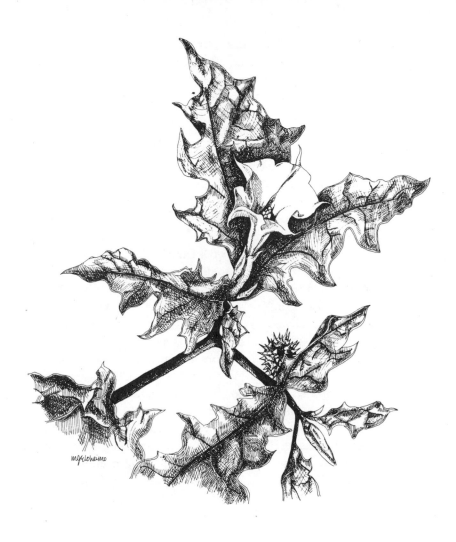

DATURA STRAMONIUM Linnaeus

BOTANICAL DESCRIPTION: An annual erect shrub, about 0.5–2.0 m in height, with either green or purple glabrous leaves and stems. Leaves 5–15 cm long, sinuately dentate and ovate. Flowers trumpet-shaped, white or lavender, erect, growing singularly from the forks of the stem. Flowers 6–10 cm long by 3–4 cm wide, corolla with 5 lobes, calyx circumcissile, five-toothed, half as long as the corolla, style shorter than the corolla. Fruit is an ovoid capsule, 6 cm long by 4 cm wide, either smooth on the surface or with spines of uniform length. Black seeds (8, 86).

HABITAT AND SEASON: This plant is cosmopolitan, and is distributed throughout the temperate regions of the world. Found especially in second-growth regions such as roadsides, dry stream-beds, and pastures. Flowers throughout the summer months (8, 86).

HALLUCINOGENIC AGENTS: Leaves, roots, seeds, and stem of this plant contain the tropane alkaloids scopolamine and atropine/hyoscyamine (159). Atropine is used in medicine, particularly in topical application to the eye, to aid in ophthalmologic examination.

HISTORY OF ETHNIC USE: *D. stramonium* is commonly called "thorn-apple," or "Jimson weed," a corruption of the appellation "Jamestown weed," which it acquired subsequent to a famous incident involving the accidental ingestion of this plant by soldiers of the colony of Jamestown, Virginia (159). In ancient Mexico, this plant, as well as other *Datura* species, was called "toloatzin," or "tolohuaxihuitl" by the Aztecs, and was used in medicine (46). *Datura* species were said to be used by the Thugis in India to drug their victims (61). These plants were also used as an admixture to *A. belladonna* in the witches' ointment (77). This plant was used as an ingredient in a medicine called "wysoccan," which was employed in adolescent rites of initiation by the Algonquins and other tribes of northeastern North America. Adolescent males of the Algonquin were confined and allowed no sustenance except "wysoccan." They were kept so for two or three weeks, during which time they were said to lose their memories of ever having been boys, thereby beginning manhood (159). *Users of this plant should exercise extreme caution—overdose may cause sickness and death.**

*See footnote, page xiii.

NICOTIANA TABACUM Linnaeus

BOTANICAL DESCRIPTION: A stout and viscid annual herb, 1 to 3 m in height. Thick erect stem with a few rapidly ascending branches. Leaves up to 50 cm in length, decurrent, ovate, elliptic or lanceolate, apex acuminate. Calyx 12 to 20 mm long, cylindric to cylindric-campanulate, viscid and unequal. Corolla 10 to 15 mm long, 2.5 to 3 mm wide, slightly curved, outer surface puberulent. Stamens erect, spaced evenly, anthers of 2 longer pairs near mouth of corolla. Capsule 15 to 20 mm long, ovoid, elliptic. Seeds spherical or elliptic, brown, about 0.5 mm in length, ridges fluted (71).

HABITAT AND SEASON: One of the oldest cultigens in North America, this plant is today of economic importance all over the world. In Pre-Columbian times, this plant was grown in the West Indies, Mexico, and Central America. It may also have been cultivated in western North America. Today this plant has escaped cultivation all over the continent (71).

HALLUCINOGENIC AGENTS: The leaves of this plant contain nicotine, an alkaloid common to all plants of the genus *Nicotiana*. While not generally considered to be an hallucinogen, this compound has potent effects on the central nervous system (59), and the ethnobotanical data warrant the inclusion of this plant in this study (61). Harman and norharman, analogs of harmine, an hallucinogenic principle of *Peganum harmala* and *Banisteriopsis* species, have been isolated from the leaves of this plant, and in higher concentrations in the smoke of combusted leaves (97, 138).

HISTORY OF ETHNIC USE: The use of this plant was first observed in the New World by Europeans in 1492, when Columbus noted that the natives of the West Indies used it, under the name "tobaco." Some years later, Sir Walter Raleigh brought specimens of a closely related plant, *N. rustica,* to the Old World. This plant was widely regarded to be an aphrodisiac, which greatly contributed to its instant popularity. Smoking and snuffing *N. rustica* and *N. tabacum* (the practices of smoking and snuffing were previously unknown in the Old World) became a fad which swept throughout Europe and Asia (61). At this time, tobacco was smoked much as marijuana is today, that is, several people would get together, roll a "joint," and each take turns inhaling the smoke, to induce hallucinosis (195). The ancient Aztecs called this plant "quauhyetl," and used it, as well as the more potent *N. rustica,* which they called "picietl" as a cleansing ablution (45, 46). It has been reported that this plant is still used by Mazatec curers, during mushroom ceremonies, for the same purpose (79). The importance of this plant to Aztec ritual is attested by the fact that a bas relief carving of a flower of *N. tabacum* appears on the Aztec statue of Xochipilli, probably the Aztec god of hallucinogenic flora, along with representations of the caps of "teonanácatl" (hallucinogenic mushrooms), flowers of "sinicuichi" (*Heimia salicifolia*), and flowers of "coaxihuitl" (*Rivea corymbosa*) (186). Chac, the ancient Mayan rain god, is often depicted blowing smoke, presumably of tobacco (61). To this day, plants of the genus *Nicotiana* are used by the shamans of the Warao and other tribes in South America, as hallucinogenic aids in the divination of the causes of illness (198). Apparently, the modern cultivated varieties are only weak hallucinogens, and are used with such frequency (tolerance is quickly acquired), in absence of any expectation of "getting high," that the modern civilized use of tobacco is not associated with altered states of consciousness.

ZYGOPHYLLACEAE

PEGANUM HARMALA Linnaeus
PLATE 11

BOTANICAL DESCRIPTION: An annual herb, with freely branching stems 20–50 cm in length, which are glabrous and glaucous, decumbent to erect. Leaflets irregular and pinnatifid, into narrowly linear segments 1–3 cm in length, stipules setaceous, approximately 1 mm in length, quickly deciduous. Flowers solitary, regular, perfect, on peduncles 1–2 cm long, 5(4) distinct linear sepals, sometimes lobed, 1–2 cm long. 5(4) distinct white petals, 14–18 mm long, with 15 stamens. Pistils 2–3, carpellary. Fruit an ovoid or globose membranous capsule, 10–15 mm long, with 3 cells and numerous seeds (86).

HABITAT AND SEASON: This is an Old World plant which has been introduced to the New World for its medicinal properties. Has escaped cultivation and grows wild in the western and southern United States and in northern Mexico. Found only in arid zones. Flowers from June through August (86).

HALLUCINOGENIC AGENTS: The seeds of this plant contain the harmala alkaloids harmine and harmaline, substituted indole alkylamines common to this, and other genera of higher plants (199). These compounds are the main active principles of yajé, or ayahuasca, which is made from *Banisteriopsis* species. Various types of yajé beverages are widely used by South American shamans (159).

HISTORY OF ETHNIC USE: It has been suggested that this plant is the "Haoma" of the *Zend-Avesta* (67). It has also been suggested that this plant represents the euphoriant plant "Soma" of the *Ṛg Veda* (67, 189). *P. harmala* may perhaps have come into use as a substitute plant for Soma in Iran. No modern use of this plant has been observed in the Old World, other than as a healing herb and dye—its use as an hallucinogen has apparently ceased (159). Recently there have been attempts to cultivate this plant in the United States, for use as a recreational drug. There is however, no information available at this time regarding use and toxicity (130)

PART TWO

The Etiology of Religion
A HISTORY OF HALLUCINOGEN USE

El honguillo viene por sí mismo, no se sabe de dónde, como el viento que viene sin saber de dónde ni porqué.

The little mushroom comes of itself, no one knows whence, like the wind that comes we know not whence nor why.

—MAZATEC saying, quoted by R. G. Wasson

THE ETIOLOGY OF RELIGION

A History of Hallucinogen Use

During the last decade, particularly in the more highly industrialized countries, there has been a phenomenal increase in the use of hallucinogenic compounds. Many people, it would seem, have come to regard this use as unprecedented, as having recent roots—a bizarre consequence of the "better living through chemistry" era. Clearly, this conception of the history of hallucinogen use is faulty and short-sighted. Indeed, there is perhaps no tradition more ancient and venerable, more common to the various human cultures—both ancient and modern, both "highly civilized" and "primitive"—than the ingestion of hallucinogenic plants.

Study of the history of this use, and its manifold implications, will afford great insight into the present popularity of hallucinogenic agents. Further, this study, by its very nature, will trace the development of religious thought from primordial times to the present, and will, it is certain, provoke in the reader truly startling insights into the possible nature of current religious practice, of the laws which seek to regulate the use of hallucinogenic compounds, and of popular, ecclesiastical, and medical opinion regarding hallucinogenic agents.

The following section details a theory regarding the origins of primitive religions in hallucinogen cults, and the subsequent evolution of these religions into modern types. The reader is referred to the remaining sections of Part Two for documentation and substantiation of this theory.

THE ETIOLOGY OF RELIGION

In pre-agricultural times, man, basically omnivorous, subsisted by hunting game and gathering wild plants. As a consequence of experimenting with plants in his environment, primitive man learned that some plants could kill, while others could nourish life. A further consequence of this experimentation with plants was the discovery of various hallucinogenic plants. Thus, in searching for food sources, particularly after migration to areas previously uninhabited by human beings, aboriginal man accidentally discovered that some plants could produce altered states of consciousness.

What a marvelous discovery that was! After the ingestion of hallucinogenic plants, our remote ancestors were able to transcend their world of mysterious and incomprehensible forces, to escape to a different state of consciousness and, consequently, a different perception of the world. Was not the concept of the "otherworld" thus born?—a conception of the world as it is experienced from the standpoint of an extraordinary way of perceiving the environment—induced by the altered states of consciousness triggered by the ingestion of hallucinogenic plants. As Wasson has expressed it:

The animal kingdom does not know God; it has no conception of the religious idea. The animal cannot imagine horizons beyond the horizon it has actually seen, a past earlier than it has experienced, a future beyond the immediate future, planes of existence other than this one in which we find ourselves. There must have come a time when man, emerging from his bestial past, first grasped these possibilities, vaguely, hesitantly; when he first knew the awe that goes with the idea of God. Perhaps these ideas came to him unaided, by the light of his dawning intelligence. I suggest to you that, as our most primitive ancestors foraged for their food, they must have come upon our psychotropic mushrooms, or perhaps other plants possessing the same property, and eaten them, and known the miracle of awe in the presence of God. This discovery must have been made on many occasions, far apart in time and space. It must have been a mighty springboard for primitive man's imagination. [Wasson, R. Gordon, *Transactions of the New York Academy of Sciences* Vol. 21(4): 333, 1959.]

While under the influence of these psychotropic plants, our ancestors saw visions, which must have symbolized the animals they hunted, the plants they consumed, the souls of their ancestors, and the unknown forces of their environment. I suggest that these visions of the otherworld came to be called "spirits" and "gods," to whose actions aboriginal man must have attributed climatic forces, astronomical phenomena, even the creation of man and other life forms.

Much of primitive man's conception of the universe was probably based on imagery derived from hallucinogenic plants. This is particularly apparent with respect to hallucinogenic mushrooms, perhaps the most abundant and cosmopolitan type of hallucinogenic plant. Our ancestors observed that mushrooms would spring forth rapidly, mysteriously, apparently without seed (spores are invisible to the naked eye). This inspired myths of the divine genesis of these magical plants, myths which are universal and persist to this day (113, 192). Ingesting these miraculous "heaven-sent" plants enabled man to commune with "god" and his world. Hallucinogenic mushrooms became archetypal mediators with the divine.

The ancient Aryans of India deified a plant that they called "Soma," which Wasson has recently identified as *Amanita muscaria,* the fly-agaric of European lore (189). The Aryans called Soma the "mainstay of the sky," a metaphor which occurs repeatedly in the hymns of the *Ṛg Veda* which refer to Soma. Evidently, the cap of the mushroom represented the vault of the sky, the stem the route to the heavens from earth—through the ingestion of the heaven-sent plant and the consequent induction of an altered state of perception.

These motifs represent a stylization of the images of the hallucinogenic plants themselves, and of the consciousness states which they produced. This stylization or ritualization was probably the work of the tribal shaman. The tribal shaman was a specialist in the use of hallucinogenic plants, and in the induction and use of altered states of consciousness. As early human cultures developed, and individual men came to specialize in specific endeavors (i.e. hunting, farming, building, etc.), the specialty of shamanism developed. The shaman was probably chosen by the people he served, who expected a prospective shaman to manifest, at an early age, a marked propensity for behavior characteristic of altered states of consciousness (58). The neophyte shaman was instructed in the use of hallucinogenic mushrooms and other plants, and his speciality was to serve as the intercessor between man and the "spirit" world. Shamans came, at an early time, to ritualize their hallucinogenic experiences. They attributed illness to bad influences from the "spirit" world, and established themselves as curers of illness—only by consulting the tribal shaman, the specialist in the dark and mysterious "otherworld," could a sick person effectively propitiate the evil "spirits" that plagued him. The shaman, then, was a "medicine man," and his function was to apply his plants—his medicine—toward curing illness. Thus a primitive theocracy came to monopolize the use of hallucinogenic plants, and created a pantheon of "spirits" or "gods" to explain various natural phenomena. This theogony and a corresponding cosmology was perpetuated by instructing succeeding generations of shamans in the ways of the old. The existence of shamanism, with its many distinguishing characteristics, has been documented in primitive cultures all over the world (58).

As human cultures became more sophisticated, the tribal shamans apparently evolved into priests. This evolution eventually entailed the cessation of the use of hallucinogenic plants. The priests served basically the same functions as the shamans, with one important difference. The ritualization eventually became so complete, that the plants originally responsible for this ritual fell into disuse, or their users were persecuted or destroyed. Following military conquest, primitive hallucinogen cults must have come to be persecuted and driven out of existence by the more advanced religion/cultures which supplanted them. The persecution of primitive cults evidently resulted in the origin of the concept of "tabu." The old "sacraments" came to be identified as evil, as relating to evil "spirit" influences, and it was declared to be a dangerous sin to partake of these plants.

Thus the shaman, who once specialized in altered states of consciousness, evolved into a priest, who preached a doctrine derived from these altered states, but who did not use the plants, who did not, in fact, have direct experience of the "otherworld." The populace, therefore, was expected to accept the priest's doctrines without proof—the beginning of "faith." In other words, upon visiting a shaman, a man could have had direct experience of altered states of consciousness, of the "otherworld," either by watching the shaman, under the influence of hallucinogenic plants, journey to the "otherworld" and return with information from the "spirits," or by himself partaking of the shaman's "medicine," his hallucinogenic plants, and accompanying the shaman to the "otherworld." When the shaman came to be replaced by a priest, however, the subject was expected to have faith, to be a "true believer," and to accept the priest's *symbolic* intercession with the "gods" on faith, without proof. Of course, some people were more skeptical than others—those who would not accept this "second-hand" religion were cast out and persecuted, and the idea of "heresy" was born. At this point, religion as a human institution began to have difficulties.

Modern attitudes toward mushrooms provide an illustration of the "tabu" concept. Owing to their early use as ritual hallucinogens, mushrooms probably came to be persecuted by the developing priesthoods, and this persecution has seemingly survived to this day. Although a very few species of mushrooms are in fact dangerous, there is a widespread belief in our culture that mushrooms as a whole are deadly. There is no greater preponderance of poisonous species among the fungi than among the higher plants, yet mushrooms are, in some western cultures, regarded with loathing and fear. It is apparent that the ancient tabus prevail. It is clear from western art that mushrooms have long been associated with evil influences, such as the "devil" (192). This fear of mushrooms, this "mycophobia," is so prominent that mushrooms are literally persecuted—in my youth I was taught that mushrooms were so many sleazy "toadstools" (toads are also associated with the "devil" in western cultures), fit to be stomped on and kicked and certainly to be avoided. I was not taught to regard flowers in this manner, although there are a number of types of flowering plants which are deadly poisonous. The point is that "myco-phobia" is a phenomenon of great interest—due to the importance of mushrooms in early religious symbolism, they have seemingly come to represent an ancient form of religion which has long since been persecuted, destroyed, and rendered "tabu." Even in this so-called "enlightened" age, this tabu survives in very common forms of inexplicable and irrational attitudes toward certain life forms.

The above theory has been presented as a concise outline of the impor-tance of hallucinogenic plants in the etiology of religious ideas, and of the consequent evolution of these ideas and the institutions which have come to be associated with them. The reader is not, however, expected to take these ideas on faith—there is an abundance of historical evidence to illustrate the above theory. The remainder of Part Two will be devoted to examining some of the most outstanding and salient features of this historical evidence, as a means of illustrating this conception of the origins of religions. Finally, in the light of this information, the present situation regarding the use of hallucinogenic plants will be placed in a perspective more useful for understanding and dealing with the problems which this use entails.

THE OLD WORLD—SHAMANISM TO SOMA

The use of hallucinogenic plants in historical times has been observed in Europe, Asia, Africa, North and South America, and Oceania (159). In almost every case where groups of people have been found living in "primitive conditions," they have been observed to be using hallucino-genic plants. The use of hallucinogenic plants, then, is a phenomenon which primitive cultures almost universally display. It must be assumed that our present highly complex cultures descend from such primitive cultures; it would not be, therefore, surprising to find parallels between modern religious motifs and practices, and those of ancient hallucinogen cults.

In an exhaustive work, *Shamanism: Archaic Techniques of Ecstasy* (58), Mircea Eliade has taken great pains to document the existence of sha-manism all over the world—both in the "primitive" cultures which have been observed in modern times, and in the historical records of modern "civilized" cultures. Not only does shamanism appear to be a common feature of human cultural evolution, wherever this has occurred, but, as

Eliade has found, there is a great deal of common symbolism and imagery inhering in virtually all of the examples of shamanism which have been observed in modern times. This probably results from the fact that all of these types of shamanism have their roots in experiences with hallucinogenic plants. The most central theme of shamanism is the idea of the "otherworld" and the shaman's ability fo "fly" to this world through the ingestion of hallucinogenic plants, and by other techniques. Another feature common to the various types of shamanism which have been observed in modern times is the idea that this "otherworld" is inhabited by "spirits" with which the shaman may commune during his ecstatic flights to this world. A striking image common to many of the instances of shamanism which have been studied is that of the "world tree," which the shaman climbs to aid him in reaching and communicating with the "spirits" of the "otherworld" (58).

It is significant that in Siberian shamanism, this "world tree" is always the birch (*Betula* spp.). In Siberia, birch is the common symbiont for the psychoactive mushroom, *Amanita muscaria*. That is, *A. muscaria* grows in a symbiotic association with birch trees, and does not grow in the absence of these trees. Less commonly pine trees (*Pinus* spp.) serve as symbionts for *A. muscaria* in Siberia. Now, in the nineteenth century, the use of *A. muscaria* as a shamanistic inebriant was widely observed in Siberia. In *Soma: Divine Mushroom of Immortality* (189), R. G. Wasson quotes several accounts of this use in Siberia:

I learned that shamans are very eager to take in a certain quantity of *Amanita muscaria* in order to get themselves into a stupor resembling complete insanity. [Dittmar, Carl von, *Historical Report*, p 524.]

The shaman must get himself into an exalted state to be able to talk to the gods. To achieve this he consumes several (either seven or fourteen or twenty-one) fly agarics [*Amanita muscaria*] which are capable of producing hallucinations. [Patkanov, Serafim K., *The Irtysh Ostyak and their Folk-Poetry*, p 121.]

A. muscaria was also widely used as a recreational drug in Siberia:

The fly agarics are dried, and eaten in large pieces without chewing them, washing them down with cold water. After about half an hour the person becomes completely intoxicated and experiences extraordinary visions. The Koryak and Yukagir are even fonder of this mushroom. So eager are they to get it that they buy it from the Russians wherever and whenever possible. Those who cannot afford the fairly high price drink the urine of those who have eaten it, whereupon they become as intoxicated, if not more so. The urine seems to be more powerful than the mushroom, and its effects may last through the fourth or the fifth man. [Steller,

Georg W., *Description of Kamchatka, Its Inhabitants, Their Customs, Names, Way of Life, and Different Habits,* p 92.]

Thus the "world tree" symbolism in Siberia is related to the psychoactive mushroom used by the shamans to achieve voyage to the "otherworld" (189). When the shaman actually scaled a "world tree," a ritual common to Siberian shamanic practice, he was probably symbolically representing the ingestion of a psychoactive plant to achieve "flight" to the "heavens." This ritual of scaling the "world tree" has been observed in areas where the shamans have ceased to use hallucinogenic plants (58) – possibly an example of the ritual surviving while the use of the plant originally responsible for the ritual does not. Interestingly, in tribes whose shamans no longer use hallucinogenic plants, the existing shamans will refer to the greater power of the ancestral shamans (58). In other words, they relate stories of the deeds of ancient shamans, who presumably actually used hallucinogenic plants and consequently experienced profound altered states of consciousness. These modern shamans believe that their existing rituals (in absence of the use of hallucinogenic plants) are only imitations, lacking the power of the ancestral rites (58).

About 4000 years ago, a people called the Aryans invaded what is now Afghanistan and northern India, from their home to the north. The Aryans established a civilization in the Indus Valley, and composed a book of 1028 hymns, called the *Rg Veda.* This is the earliest of a group of four Indian holy books, collectively called the "Vedas," and its writing represents the inception of the so-called "Vedic Age" in India. Many of the hymns of the *Rg Veda* extol the virtues of a plant called "Soma," which was deified by the ancient Aryans, and was undoubtedly hallucinogenic. Long a subject of ethnobotanical controversy, Soma has recently been identified as *Amanita muscaria* by R. G. Wasson (189). The main arguments in favor of this identification are: 1) the *Rg Veda* states that Soma grew in the mountains, as does *A. muscaria* in the area in question; 2) although the *Rg Veda* describes Soma in vivid, colorful terms, it makes no mention of leaves, roots, seeds, or flowers, though it does state that the plant was fleshy and juicy – *A. muscaria,* being a mushroom, of course has no leaves, roots, seeds, or flowers; and is indeed fleshy and juicy; 3) the *Rg Veda* describes Soma as both a plant and the urine of a priest who had ingested the plant:

Acting in full concert, those charged with the Office, richly gifted, do full honor to Soma. The swollen men piss the flowing [Soma]. [*Rg Veda* Maṇḍala IX 74[4], quoted by R. G. Wasson.]

There are, then, compelling arguments in favor of the identification of Soma as *A. muscaria*. Not only did the Aryan practice of the ingestion of Soma and Soma-urine come from the north (i.e. the direction of Siberia), but a similar *and unique* practice (never observed in any other area), that is, of ingesting the urine of the user of an hallucinogenic plant, has been observed to exist in modern times near the area from which the Aryans were said to have come (55, 189, 190).

Today, although the *Rg Veda* hymns dealing with Soma are still sung in the rituals of modern Hinduism, the actual Soma plant is no longer in use. Moreover, it is evident that, in the intervening two millennia since Soma ceased to be used, various substitute plants have been employed in the Indian rituals (189). This example, then, illustrates the concept of the evolution of religions. The ceremonial ingestion of Soma apparently originated in primitive shamanic practice. The use of this hallucinogenic plant came to be dominated by an organized priesthood. The rituals involving the use of this plant were written down in the *Rg Veda,* which became the foundation of the modern religion, Hinduism. While the theocracy and priesthood survives to this day in India, and uses the original Soma hymns in rituals, the actual Soma plant, originally responsible for these rituals, is no longer used in conjunction with them.

Further, it is significant that the modern Hindu religion is characterized by techniques collectively known as yoga, which are essentially designed to produce altered states of consciousness, without the use of plants or any other external influence. These yoga techniques—sensory deprivation, breath control, meditation, fasting—are common to modern Hinduism, Buddhism, and Zen, and are known to result in altered states of consciousness. It is interesting to note that the first known text devoted to descriptions of these techniques, the *Yoga Sutras of Patanjali,* also lists the use of drugs as a means to attain the powers of yoga (194), although modern practicioners consider drug use to be detrimental.* Yoga techniques, therefore, probably represent a system for producing altered states of consciousness, without the use of the plants which were responsible for the cultural tradition of producing altered states of consciousness in religious ritual. Part Three of this book deals with mechanisms for the induction of altered states of consciousness by both drugs and yoga techniques.

Eliade has documented the existence, in Siberia, of various other shamanic techniques for the induction of ecstatic trances, without the use of hallucinogenic plants. Siberian shamans often employ drumming (on drums made of birch), singing and chanting, and other techniques to produce ecstatic states (58). These techniques are curiously similar to some aspects of yoga practice, which also employ rhythmic chanting and singing. There are many other examples of non-chemical techniques for the induction of ecstatic states. In the Middle East, Sufi "whirling dervishes" were known to dance and spin around to achieve trance-like states. In the United States, Christian worshippers still hold "revival meetings," during which the participants will sing and chant and pray, in hopes of producing possession by the "spirit of the lord." As is the case in Siberian shamanism, modern participants in non-chemical ritual techniques for the induction of ecstatic states have traditions of ancestral "saints" and "avatars" supposedly endued with more power, more "enlightenment" or "grace of god" than the present day priests and practitioners.** It is commonly believed that only if the "faith" of the modern practitioner is sufficiently strong, can he hope to attain to the "true power." These modern ways of producing ecstatic states may represent attempts to induce archetypal "religious states" without the use of the hallucinogenic plants originally responsible for the tradition of ritual alterations in consciousness.

THE NEW WORLD—TEONANÁCATL AND THE BODY OF CHRIST

Thousands of years before the establishment of the Soma cult in the Indus Valley, bands of nomads migrated from Siberia into North America, across a land bridge over what is now the Bering Straits. This land bridge has long since subsided, separating the so-called New World from the Old World. Over a period of millennia, these nomads have migrated throughout the American continents, eventually colonizing Patagonia, at the southernmost extreme of South America. Along with other common shamanic traditions, these nomads brought the practice of ingesting hallucinogenic mushrooms and other hallucinogenic plants to the New World.

In historical times, the use of hallucinogenic mushrooms has been

---

*In a modern edition of the *Yoga Sutras,* the translators comment on Patanjali's reference that certain drugs may have negative effects on spiritual development and belief and may even cause permanent brain damage. See: S. Prabhavananda and C. Isherwood, *How to Know God—The Yoga Aphorisms of Patanjali,* Vedanta Press, Hollywood, 1971.

**The intention here is neither to denigrate these traditions, nor to suggest that "saints" and "avatars" are fakes. See Part III, "Notes on Hallucinosis Beyond the Drug State."

observed in Mexico and Peru (79, 159). Use of narcotic fungi has recently been observed among certain Alaskan tribes (146). There has been abundant documentation of the use of other hallucinogenic plants throughout North and South America (159). Part One of this book deals with some specific instances of this use in North America. The most widespread use of hallucinogenic mushrooms has been found in Mexico, principally among the Mazatec, Chatino, and Zapotec Indians of the state of Oaxaca, and among the Nahua Indians of the state of Mexico (79, 192).

Examination of historical records shows that the ancient Aztecs of Mexico used hallucinogenic mushrooms and other hallucinogenic plants. A number of documents from the time of the conquest of Mexico attest to this fact (79, 159, 192). Hernández, personal physician to the king of Spain, described the use of mushrooms called "teonanácatl" or "teyhuinti" which:

... cause not death but madness that on occasion is lasting ... There are others that ... bring before the eyes all sorts of things, such as wars and the likeness of demons. [Hernández, H., *Nova Plantarum, Animalium et Mineralium Mexicanorum Historia.*]

Other writers described this practice:

... and that they had eaten nanacates [i.e. teonanácatl, Nahuatl name for magic mushrooms] to invoke the devil as their forefathers had done." [Don Diego, *Codex of Yanhuitlán,* translated by the author.]

Archaeological work in Mexico has indicated the antiquity of ritual mushroom use. Pre-Toltec frescoes at Teotihuacan show an association between mushrooms and the Toltec rain god, Tláloc, and an association between mushrooms and the "otherworld" (79, 192). These frescoes have been dated at about 300 A.D. Some of the oldest known archaeological relics of Mexico and Central America are "mushrooms stones"—stone mushrooms with idols carved into the stems (111, 192). A drawing of one such icon appears in the front of this book. The oldest of these stones have been dated at around 1000 B.C. The Aztec statue of Xochipilli, the "Prince of Flowers," which is currently displayed in the National Museum of Anthropology in Mexico City, bears carvings of the caps of mushrooms, as well as the flowers of other hallucinogenic plants. This statue was found at Tlalmanalco. Xochipilli was probably the god of hallucinogenic flora in the Aztec pantheon (186). Thus it is apparent that in the New World, there is a very ancient tradition of the use of hallucinogenic mushrooms and other psychotropic plants.* Further, there is abundant documentation of the ritual use of hallucinogenic mushrooms and other magic plants in modern Mexico (79, 159).

When the Spaniards invaded Mexico in the early sixteenth century, they felt an obligation, owing to the teachings of their religion, Catholicism, to destroy the so-called "pagan" practices of the primitive Indians and to supplant them with Christian practices. Accordingly, they embarked on a bloody course of conquest and conversion, which resulted in the subjugation of Mexico, or the Aztec empire, in 1521, and the eventual subjugation of the few remaining tribes in remote areas of Mexico during the next century and a half (141).

The above quotations indicate that the Spaniards associated the ingestion of mushrooms in religious ritual with the worship of "demons" or the "devil." It is apparent from European art of the middle ages that mushrooms themselves were associated with the "devil"–the Christian "god" or "spirit" entity which embodied evil influences. There are indications of associations between mushrooms and the "devil" in Spanish art from Mexico at the time of the conquest (84, 192). The Spaniards came from a "mycophobic" culture, and they were aghast at the use of mushrooms in religious ritual in Mexico. Not only were the Spaniards "mycophobic," but the *Bible,* their religious book, specified that other, allegedly misguided persons would worship "devils" and that it was, as a consequence, the duty of a "good Christian" to use whatever means were necessary to make such misdirected persons see the error of their ways. This teaching in the *Bible* seemingly represents the written programming of the persecution of the primitive religions which preceded the Biblical religions. With these facts in mind, it is hardly surprising that the Spanish invaders so vehemently persecuted the religious cults of the New World–they saw in these cults the verification of a certain set of their religious teachings, and felt that it was their duty to destroy these "pagan" cults. Of course, the fact that the Spaniards were interested in acquiring wealth, and that the Mexicans possessed gold and other valuables made this seem only more logical. Certainly, the religious zeal of the invaders served as a very thin pretext to

*It has been suggested that page 24 of the Mixtec *Codex Vindobonensis* depicts a number of deities holding mushrooms, and the youthful god-king Quetzalcoatl talking to a goddess that the mushrooms have apparently invoked. This codex was painted in the sixteenth century, and is one of the best preserved and most artfully executed of the Mexican codices. For details, see: A. Caso, "Representaciones de Hongos en los Códices," *Estudios de Cultura Nahuatl,* Vol. IV, Universidad Nacional Autónoma de México, 1963.

conceal another motivation—greed. Modern scholars cringe at the accounts of the Spaniards spending whole days destroying "idols," priceless remnants of past civilizations. On one occasion, in 1561, Bishop Diego de Landa burned a stack of all of the ancient Mayan codices which he could find, because of his abhorrence at the religious rites which they documented (103). Landa wrote:

Among the Maya we found a great number of books, written with their characters, and because they contained nothing but superstitions, and falsehoods about the Devil, we burned them all, which the Indians felt most deeply, and over which they showed much sorrow. [Landa, D., *Relación de las Cosas de Yucatán.*]

Only three authenticated Mayan codices survived this, and presumably other conflagrations. Although these books have not been fully deciphered, it is apparent that they were written to document the religious rituals of the ancient Mayan civilization. It is interesting to note that two of these three surviving codices show representations of what appear to be mushrooms, drawn in association with figures of priests and animal deities (112).

This behavior on the part of the Spaniards demonstrates how extreme was their loathing for the ritual practices of the Mexican Indians, characterized as they were by the ingestion of "toadstools," or "devil's bread," as the hated mushrooms were known in Europe. Of course, the association between the use of mushrooms and such hideous practices as human sacrifice only served to deepen the disgust of the Spaniards (141).

This, then, is an example of "mycophobia," and the intensity of the Spanish reaction to the use of mushrooms testifies to the influence of the old "tabus." There is no evidence that any of the Spanish conquerors tried the mushrooms, or attempted to study them. The Mexican/Spanish interaction is a classic example of the persecution and destruction of hallucinogen cults following military conquest. Mexico today is a Catholic country, and the use of hallucinogenic fungi in indigenous rituals has been observed in only a few isolated areas. In every case where this use has been documented, the rituals have been found to be inextricably intermingled with imagery, ritual, and dogma from the Catholic religion (163, 188, 192).

The persecution of the Mexican "tabu-breakers" is not without historical parallel. At the same time in Europe, the authorities of the Christian religions were engaged in a two centuries old war against so-called "witchcraft," apparently a last vestige of the primordial hallucinogen cults which must once have existed in Europe. The "witches" were noted for using hallucinogenic plants (77), and were said to worship "devils." In this case, too, the Christian teachings regarding the worship of "devils" served as an excuse for barbarous and execrable behavior—the torture and execution of thousands of "heretics" or "tabu-breakers." Particularly in Spain, much of the medieval Inquisition was directed against Jews. Of course, Judaism is the father religion of Christianity, Christ himself allegedly having been a Jew. We are thus confronted again with an example of the persecution of old religions by the exponents of the newer religions which had evolved from them. The persecution of Jews survives to this day, and is likely to last as long as Judaism itself.

The persecution of "witches" in Europe resulted, in part, from the fact that some of the "witches" were known to use plants in the treatment of illness (77, 172). Of course, some contemporary orthodox physicians also prescribed plants for the treatment of disease. The "witches," however, attributed their cures to "magic" or the forces of the "devil," or some such "spirit" influence, not in keeping with the orthodox opinion that only "god" could cure illness, albeit through the agency of the plants he had created for man's use. Jules Michelet states:

[the Church] . . . declares, in the fourteenth century, that if a woman dare cure *without having studied* [i.e. the Scriptures], she is a witch and must die. [Michelet, J., *Satanism and Witchcraft,* p xix.]

Not only, then, did the European "witches" and the Mexican Indians break the old "tabus" by using hallucinogenic plants, but they defied the authority of the established religions by curing illness and attributing the cures to "pagan" deities, that is, to "spirits" other than those (the virgin Mary, Jesus, etc.) accepted by the establishment.

Immediately following the conquest of each respective tribe in Mexico, the Spaniards attempted to convert the vanquished natives to the Catholic faith. As a rule, the Mexican Indians were impressed with the Spaniards, even in some instances believing them to be "gods." It is evident that many of the native groups were awed by the splendor of the rituals which the Spanish priests performed for their benefit (141). They had never before seen such finery as gilded crosses and altarpieces, velvet cloths, and so forth. Clearly, this was, in the eyes of many of these Indians, more splendid and grander than the indigenous rituals. The Spaniards offered communion to the natives, explaining that the wafers and wine, or "holy sacrament" represented the flesh and blood of their "god." Now, the Mexican name for hallucinogenic mushrooms was "teonanácatl," which means "god's flesh," and, upon hearing this explanation of the Christian communion, the Indians surely expected to receive an hallucinogenic substance like their own "god's flesh." It is probable, moreover, that they expected an hallucinogenic experience that much grander than their own, in proportion as the Spanish rituals were grander than the Mexican. Indeed, it is difficult to imagine them entertaining any other expectation. Coming as they did, from a *primary* religion, a religion characterized by direct experience of the "otherworld," how could these Indians have had any conception of any other type of religion, some such religion as depended on *symbolic* communion, on faith, without the actual use of an hallucinogenic substance? Furthermore, it is evident that many of the Mexican Indians didn't "get off" on the white man's religion, and they attempted to return to their own, which allowed direct experience of altered states of consciousness (141).

The Christian ritual of communion itself may represent the symbolic ingestion of hallucinogenic plants. As a consequence of the evolution of religions, the original plant which would have been responsible for this ritual is itself no longer used, but the ritual survives. This is directly analogous to the survival of the Soma rituals in India, in absence of the use of Soma itself. This is but one example of a modern religious motif which suggests an ancient tradition of use of hallucinogenic plants. Another motif common to modern religions is that of the tree of learning, knowledge, or enlightenment. This is, possibly, a representation of an archetypal hallucinogenic plant.* In Biblical lore, Adam and Eve were forbidden to eat from the "tree of knowledge"–this represents another example of the concept of "tabu." Had Adam and Eve eaten of this supposedly hallucinogenic plant, the *Bible* states, they would have been as "gods," that is, they would have had direct experience of the "otherworld." This was "tabu" or forbidden. It is interesting to note that Eve was tempted to eat this supposed hallucinogenic plant by the "devil"–if this interpretation is correct, then we have another example of the association of hallucinogenic plants with evil "spirit" influences. The "original sin," then, may have been to ingest an hallucinogenic plant which had been, owing to the evolution of religious thought, rendered "tabu." Those who had written the *Old Testament* established a hierarchy of priests, serving as the intercessors between man and the "gods"–the old ways respecting direct intercourse

*This may be an oversimplification. As mentioned earlier, in Siberia the "World Tree" or "Tree of Life" is the birch. R. G. Wasson has suggested that the tree of learning, knowledge or enlightenment is a representation of the birch itself, while *A. muscaria*, growing in a mycorrhizal association with the birch, represents the "fruit" of the tree of knowledge. It has been shown that primitive peoples in Siberia were aware of the fact that *A. muscaria* grew in an association with birch trees. For details, see references 189 and 192 in the bibliography.

with the "gods" by man, through the ingestion of hallucinogenic plants, were made "tabu" or sinful.

There are other motifs in modern Christianity which suggest relation to ancient hallucinogen cults. "Holy water" in Christian ritual may represent the second form of Soma—the urine of the priest who had ingested the hallucinogenic plant. This water was made "holy" by the priest, that is, as a consequence of passing through the priest's body, the water was "sanctified" and had the potential to induce communion with the "gods" (189). In *The Sacred Mushroom and the Cross,* John Allegro has suggested, on linguistic grounds, that the Christian concept of the virgin birth may represent the growth of a sacred mushroom without seed, the mushroom so created being the "son" who represented the "way" to the "father" (5).

The lotus is a common motif in modern Buddhism and Hinduism. The lotus and related plants (*Nymphaeaceae*) are psychoactive and may have been used in ancient times in connection with religious rituals, although no such use survives to this day (52). Hindu and Buddhist deities are often depicted in meditation (a technique used to achieve "enlightenment") sitting atop stylized lotus plants. Further, the "sahasrara chakra," the seventh and highest meditation or energy center of Kundalini yoga is depicted as a "thousand petalled lotus," and its locus in the body is at the top of the head. The lotus, or water lily motif also occurs in ancient Mayan art in Mexico (52). Recent work at the University of Mexico has suggested the existence of psychoactive water lilies in the Maya area (49, 130).

Other modern examples of ancient hallucinogen motifs in pseudo-religious contexts can be cited. The common Halloween image of a "witch" riding on a broomstick probably relates to an ancient mode of ingesting hallucinogenic plants. The European "witches" would prepare an ointment of hallucinogenic plants (see Part One), which was either applied directly to the body, or to a broom handle. The broom handle was then used for masturbation by female "witches," and the ointment was absorbed through the mucous membranes of the rectum and vagina. The ointment was said to induce a sensation of flight—thus the idea of "riding" the broomstick and "flying" came to be associated with "witches" (77). Santa Claus may also be an ancient image relating to the ingestion of hallucinogenic plants. Like Santa Claus, the ancient Siberian shamans possessed reindeer, lived in the North, and were known to enter the homes of people through the smokehole, to perform magic rituals (146). If this interpretation is correct, it is indeed ironic that such an ancient symbol came, indirectly, to be associated with a modern religion.

It is apparent, then, that modern religious practices show evidence of their heritage of a tradition of ceremonial ingestion of hallucinogenic plants. In the case of the Hindu religion, the Soma plant has not been used for some two millennia. Whatever plant may have been responsible for the hallucinogen motifs in modern Christianity must have, likewise, ceased to be used over 2000 years ago. It is not surprising that, being deprived of the experiences (i.e. with hallucinogenic plants) which, it is assumed, account for much of modern religious dogma, people began to lose interest and faith in these religions. Indeed, very few persons in the United States today believe in the teachings of the modern religions, no doubt because of the fact that many of these people have had no direct experiences with profound alterations in consciousness. Further, modern scientific investigations have invalidated many of the explanations of natural phenomena which grew out of primitive conceptions of the "otherworld."

THE HALLUCINOGENIC RENAISSANCE

Today, despite the evolution of religions and the resulting persecution of the users of psychotropic plants, hallucinogenic agents have once again become accessible to the people. Never before, in fact, has such a variety of hallucinogenic compounds been available to such a large group of people. It seems moreover, that having grown weary of "second-hand religion," many people have come to desire a more direct form of communion with the "otherworld," and are unwilling blindly to accept dogma. Not surprisingly, after experiencing altered states of consciousness induced by hallucinogenic agents, many people gain a greater understanding of religious ideas, and a renewed interest in pursuing them. Thus we are seeing a resurgence in interest in Christianity and Oriental religions, on the part of present and former users of hallucinogenic drugs, for the ingestion of hallucinogenic agents enables one to experience the probable source of religious ideas and imagery. Of course, the churches invariably denounce the use of these substances, although they may not really recognize this use as "original sin" or "tabu-breakage." This denunciation is, however, lacking in consequence when compared with such phenomena as the medieval Inquisition and the destruction of Mexican hallucinogen cults, for the role of the churches has been usurped by their apparent successors in the evolution of religions—modern medical science, and the other sciences peripheral to it. It would seem that modern medicine has become the religion of the day, the doctor its priest. Although there has been a resurgence, on the part of hallucinogen users, in interest in religious ideas and practices, the official power of organized religions in our society has been, in recent times, destroyed with amazing speed. The "infallible" dogma of the scientific method has replaced the Church as the "right arm" of the State (172).

Curiously, we have come full circle. Although the great Western religions had effectively eliminated the use of hallucinogenic drugs, by making these drugs "tabu" and persecuting the users, and although this prohibition of these substances has been enforced for centuries, the advent of the modern chemical age has radically changed this situation. Drugs and drug use have seemingly *become* the dominant "religion," and the modern preoccupation with pharmacological agents has led to a resurgence in interest in the old "tabu" substances. Accordingly, these substances have been singled out by modern "priests," identified as harmful, and users of these substances have been attacked as viciously and relentlessly as were the European "witches." Thomas Szasz has pointed out this parallel:

The result of the medieval rivalry between witches and priests was the Inquisition, with all its complex and far-reaching consequences—in particular the development of a powerful group of witch mongers, whose vested interest it became to produce ever more witches in order to make themselves ever more indispensable and wealthy. The modern persecution of drug abusers and pushers has generated a similar Medical Inquisition, with complex and far-reaching consequences—among them, the development of a powerful group of addiction mongers whose vested interest it is to produce ever more drug abusers in order to make themselves ever more indispensable and wealthy. [Szasz, T., *Ceremonial Chemistry,* p 66.]

Today, then, in place of religious dogma and ritual, we have "current medical opinion" and "therapy." In the place of "witches" and "heretics," we have "addicts," "pushers," and "drug abusers." The modern religion is one of health, and its sacraments are "therapeutic agents." The drugs which are not accepted by the medical "priests" are identified as "dangerous," "addicting," and "toxic" (172). Szasz further illustrates this phenomenon:

Thus, just as in religions there are good and bad, benevolent and malevolent deities—so in the secular religion of our drug age there are good and bad, therapeutic and toxic drugs. The adherents to our majority religions thus congregate at cocktail parties and "smokers," and have elaborate ceremonies symbolizing the virtues of mixed drinks and wines, cigars and cigarettes, pipes and tobaccos, and so forth. These are the holy communions of our age. Those who reject the doctrines of our principal religions and who cultivate instead various heretical faiths, congregate at pot and acid parties and at gatherings where heroin and other even more esoteric and forbidden drugs are used; and they too have elaborate ceremonies symbolizing the counter-virtues of marijuana and LSD, incense and Oriental mysticism, and so forth. These are the unholy communions of our age. [Szasz, T., *Ceremonial Chemistry,* p 41.]

The "priest" in the white coat makes all the difference. When administered by a medical "priest," morphine is regarded to be "therapeutic." The "priest" who uses this and other drugs is considered to be a saint and a savior, and our society rewards him with the utmost in respect and an abundance of material wealth. Morphine sold illegally on the street, however, is considered to be deadly and poisonous, "the most dangerous drug we know," and the man who sells it, if we are to believe our leaders and our medical "priests," is so much vile vermin worthy of extermination.

A highly publicized case in Mexico further elaborates this point. Salvador Roquet, a Mexican psychiatrist, was recently arrested for administering illegal hallucinogenic drugs to his patients. Despite horror stories in the press, and an investigation by the legislative branch of the Mexican government, it became apparent that Roquet's patients were convinced that the drugs were therapeutic and beneficial. Not a single patient would come forth and testify against Roquet. Orthodox "priests" of medicine in Mexico and the United States consider hallucinogenic drugs to be therapeutically worthless, to be, in fact, "toxic" and "dangerous." Roquet, the "heretic" who dared to defy the "dogma" of the established "medical religion," meanwhile sat in prison, another "tabu-breaker" in the stocks (130).

The medical "priest" who "sanctifies" drug use by writing a prescription, makes all the difference indeed. Thus the patient slyly savors the effects of his morphine prescription, or, in the case of Roquet's patients, his "medicinal" dose of LSD, a reward for paying obeisance to the physician "priests"—all the while contemplating with puritanical horror those degraded, unclean, and unholy "junkies" and "acid freaks," and their evil, savage "drug kicks."

Yes, we have come full circle. In primordial times, the physician was the priest, the shaman or "medicine man," and his "medicine" was hallucinogenic plants. Through the agency of the altered states of consciousness induced by his "medicine," the primitive shaman was able to divine the causes of illness. As he became more sophisticated, the tribal shaman learned to use a large number of plants, and his "medicine" expanded from diagnosis to embrace therapy. Gradually, over centuries, the offices of priest and physician evidently diverged into two separate functions, although these retained close ties until relatively recent times. The purely ecclesiastical priests, however, by virtue of the fact that they had ceased to use plants, had abandoned their "medicine," the true source of their power. The medical "priests," who still used "medicine" became increasingly stronger, finally surpassing the ecclesiastical priests in power and influence.

The discovery of the structure of DNA in 1953, and the consequent death of the last beliefs in "vitalism" marked this turning point. At this time, it came to be believed and accepted that cells and bodies obeyed chemical laws, which were supposed to be subject to the manipulations of the physician "priests." Henceforth, the ecclesiastical priests had to be careful not to infringe on the "dogma" of "medical evidence." They had to accept the notion that such things as microbes caused disease, that "medicines" alone caused cure, and the old "gods" simply weren't involved at all. After all, hadn't it been "proven" that man descended from apes, that the forces of disease could be controlled by mere vaccines?

The physician "priests," however, had also abandoned the original source of their power. They had forsaken the divinatory plants for strictly curative agents. It was only a matter of time, in this age of chemistry, before the common people again discovered the hallucinogenic, divinatory plants, and the explosive nature of this rediscovery is only now being felt. The medical "priests" are taking their turn, attempting to persecute the "heretical" cults, but this battle is quickly being lost. Medicine is returning to the hands of the people, where it began, and where it probably belongs.

Far from being a passing fad, the modern use of hallucinogenic agents in western cultures continues to grow and is here to stay. It is apparent that the temporary altered states of consciousness produced by hallucinogenic agents have been instrumental in permanently changing the consciousness of an entire generation. Those persons involved in the hallucinogenic renaissance are turning away from organized medicine, embracing instead the use of healing plants and dietary control, natural childbirth, astrology and other, even more arcane disciplines. The heightened self-awareness implicit in the hallucinogenic experience must be the mechanism of this startling atavistic trend. Already, we are seeing the rise of the modern American shaman, born, like the Phoenix, from ashes—the ashes of a tradition long dead in western culture. The spreading use of hallucinogenic agents may signify the impending death of scientific medicine —the people are experiencing the "truth"... and the "truth" is setting them free.

POSTSCRIPTUM

The Garden of Love

I went to the Garden of Love,
And saw what I had never seen:
A Chapel was built in the midst,
Where I used to play on the green.

. . . .

And the gates of this Chapel were shut,
And "Thou shalt not" writ over the door;
So I turn'd to the Garden of Love
That so many sweet flowers bore;

. . . .

And I saw it was filled with graves,
And tomb-stones where flowers should be;
And Priests in black gowns were walking their rounds,
And binding with briars my joys & desires.

—WILLIAM BLAKE, *Songs of Experience*

102

PART THREE

The Biochemistry of Emotion

Kick is seeing things from a special angle. Kick is momentary freedom from the claims of the aging, cautious, nagging, frightened flesh.

—WILLIAM BURROUGHS, *Junkie*

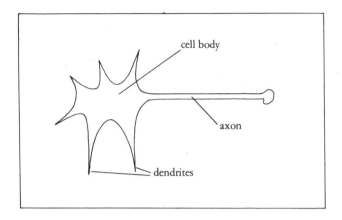

FIGURE 1. NEURON

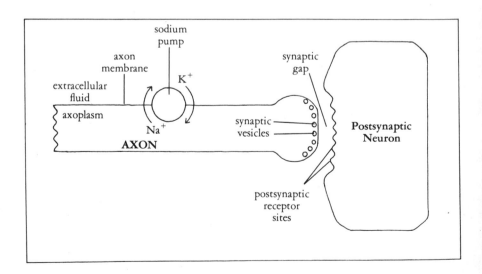

FIGURE 2. AXON AND SYNAPSE

THE BIOCHEMISTRY OF EMOTION

PHYSIOLOGY OF NEURONS

The vertebrate nerve cell, or neuron, consists of a cell body from which numerous processes extend. These processes are called dendrites, with the exception of the largest, specialized process called the axon (see Figure 1). Nerve impulses, or action impulses, are conducted by the axon of the neuron, propagating from the cell body toward the synapse, a narrow gap between the axon terminus and the dendrite or cell body of the adjacent neuron on which it impinges. While the cell body is very small, the axon is often quite large, in the human body exceeding perhaps 18 inches in length in some cases. Transmission of nerve impulses is not simply analogous to flow of electricity in a wire; actually, the cable property, or conduction, of the axon is quite low, biological material being in general a good insulator (100).

Transmission of nerve impulses is a function of depolarization of the axon membrane. That is, the nerve impulse represents a change in the charge distribution, or potential, across the membrane. The resting potential of the axon membrane is approximately -60 mv, such potential being due to the differential distribution of ions (charged particles) across the membrane. In general, there is a high concentration of potassium ions inside the axon, and a high concentration of sodium ions in the extracellular fluid. There are relatively low concentrations of sodium ions inside the axon and potassium ions without—to establish equilibrium, sodium ions tend to diffuse into the axon while potassium ions tend to diffuse to the outside. There is an active transport system for these ions in the axon membrane—the "sodium pump"—this enzyme-like protein uses metabolic energy to exchange one sodium ion from the axoplasm with one potassium ion from the extracellular fluid, both ionic transports being against the respective potential gradients (see Figure 2). In other words, the membrane performs work, to push ions across this barrier, in the opposite

105

direction in which they would normally move. In this manner, the resting potential of –60 mv is maintained, in spite of ionic diffusion. The action potential, or nerve impulse, has a value of approximately +40 mv, and is caused by a sudden influx of sodium ions into the region of the axon being measured. This is a local phenomenon which spreads unilaterally down the axon away from the cell body. Thus the nerve impulse is seen as a wave of depolarization, or change in charge distribution, propagating down the axon in response to some membrane change which allows rapid influx of sodium ions. This membrane change is a rapid phenomenon, within 2 msec (2/1000 second) the membrane regains its resting properties, and the metabolic pump rapidly extrudes the excess sodium ions which have entered the axon, exchanging them with the potassium ions which have leaked out, restoring the resting potential. This membrane change is doubtless an allosteric effect—as the membrane opens "sodium pores" at the axon hillock (the region of the axon closest to the cell body), sodium rushes in, initiating the impulse. This act of stretching open pores on one section of the axon membrane apparently induces the pores on adjacent segments of the axon to be similarly stretched open, allowing the influx of sodium ions, and hence the impulse, to propagate down the axon. Each section in turn regains its resting configuration due to elastic properties, the pores close, and the sodium pump restores the resting potential. Our primary concern in this discussion is what happens at the synapse—the mechanism of transfer of an action impulse from one axon to the adjacent neuron, across the synaptic cleft, in absence of continuity of the axon membrane (100).

The impulse is conducted across the synaptic gap by chemical agents called neurohumors, or neurotransmitters. When the action impulse reaches the axon terminus, it causes the axon to secrete discrete packets of neurohumors, which then diffuse across the synaptic cleft and are subsequently bound to specific receptor sites on the membrane of the post-synaptic (i.e. after the synapse) neuron (see Figure 2). Generally, these neurohumors are bound in synaptic vesicles lying close to the presynaptic (i.e. before the synapse) membrane; the allosteric membrane change producing impulse conduction apparently causes these vesicles to discharge their contents into the synaptic cleft. The act of binding neurohumors on the postsynaptic membrane evidently induces the membrane change which is responsible for rapid influx of sodium ions, such change in the membrane and its potential being propagated down the membrane of the postsynaptic neuron in the manner described. It must be mentioned that the binding of neurohumors may also result in the inhibition of impulse conduction by the postsynaptic neuron. Further, it may take several impulses from the presynaptic neuron (or neurons) to cumulatively induce the postsynaptic neuron to "fire," or produce an action impulse. Subsequent to binding and release from the postsynaptic membrane, the neurohumors are reabsorbed by the presynaptic neuron, where they may be degraded or reassimilated by the synaptic vesicles for reuse (100).

VERTEBRATE NERVOUS SYSTEMS

Nervous control in higher vertebrates is mediated by three systems: the central nervous system, consisting of the brain and spinal cord, the peripheral nervous system, consisting of the motor (controlling the muscles) and sensory nerves, and the autonomic system, having two components—the sympathetic and parasympathetic nervous systems. The autonomic nervous system, as the name implies, is functionally autonomous. It is responsible for homeostasis—the maintenance of blood pressure, body temperature, breathing rhythm—controls the digestion of food and serves to adapt the organism to stressful and dangerous situations. The autonomic nervous system functions without the control or intervention of the conscious apparatus—it is in this sense that it is considered autonomous.

It is clear that the neurohumors of the autonomic nervous system are acetyl-choline (ACh) and norepinephrine (NE), the neurons sensitive to these compounds being called respectively cholinergic (from choline) and adrenergic (another name for NE is noradrenaline). These two compounds have been shown to have excitatory effects on some central nervous system neurons, and inhibitory effects on other central neurons. Similarly, serotonin, or 5-hydroxy tryptamine (5-HT) has been shown to excite some central nervous system neurons, while inhibiting others. No less than eight compounds have been identified in vertebrate brains with putative neurotransmitter activity in vitro (i.e. in glass—in artificial conditions): DOPAmine, histamine, γ-aminobutyric acid (GABA), glutamic acid, glycine, as well as NE, ACh, and 5-HT. All of these substances, with the exception of GABA and glutamic acid, are bound in presynaptic vesicles. Clearly, we are confronted with a very complex situation, which will not admit of easy comprehension—a human brain consisting of some 13 billion neurons, which can be affected by at least eight different neurohumors, each of which can potentially induce or inhibit neurotransmission. Further, it is probable that there is some interaction between neurohumors, whether cumulative, competitive, or synergistic—it is no wonder that the experimental data regarding the mechanisms of drug action often appear to be confusing and contradictory. Nevertheless, much work has been done—the mechanism of NE mediated synaptic transmission has been well researched; it will be instructive to examine this in detail (100).

BIOCHEMISTRY OF NE NEUROTRANSMISSION

NE is synthesized from the amino acid tyrosine (derived from proteins in the diet) in the axoplasm of adrenergic neurons. Tyrosine is hydroxylated by the enzyme tyrosine hydroxylase to form L-dihydroxyphenylalanine (L-DOPA), converted to DOPAmine by the enzyme DOPA decarboxylase, which is subsequently oxidized by the enzyme DOPAmine-β-hydroxylase to NE. Some of the NE so synthesized is degraded intracellularly (that is, within the cell) by the enzyme monoamine oxidase (MAO) to 3, 4-dihydroxymandelic acid, which is excreted in the urine. The first two steps of this synthetic pathway occur in the axoplasm; the presynaptic storage vesicles take up DOPAmine from the axoplasm by active transport, dependent on ATP and magnesium ions. It is within these vesicles that DOPAmine is oxidized to NE. Although some of the NE so synthesized exists in the free state (as some diffuses back to the axoplasm to be oxidized by MAO), in storage it is apparently in a bound or inactive state. When the neuron "fires," the vesicles apparently discharge NE into the synaptic cleft in its free, unbound state. The NE so released then diffuses across the synaptic gap, to be bound by specific receptor sites on the postsynaptic membrane, there inducing some membrane change which will result in the initiation of a new impulse, or in the inhibition of impulse activity. The evidence indicates that much of the NE that is released into the synaptic cleft is reabsorbed by the presynaptic neuron, where it may be either degraded by MAO or taken up into storage vesicles for reuse. Any NE which diffuses into the synaptic gap and is not subsequently reabsorbed by the presynaptic neuron is enzymatically degraded to normetanephrine by catechol-O-methyl transferase, which is circulated in the blood stream, to be degraded by MAO in liver and kidney to 3-methoxy-4-hydroxymandelic acid, which is excreted in the urine (56, 197).

The mechanisms for the neurotransmitter activity of the other neurohumors discussed, with the exception of GABA and ACh, are either unknown or poorly understood. At this stage, it can only be assumed that glycine, histamine, 5-HT, DOPAmine, and glutamic acid—if indeed these compounds function as neurohumors *in vivo* (that is, in actual physiological conditions)—transmit impulses across the synapse in a manner analogous to NE. To avoid confusion, this discussion will employ the NE mechanism as an example of synaptic neurotransmission, and in discussing the effects of the various drugs on neurohumors, it will be assumed that all of these compounds mediate synaptic transmission by functionally analogous mechanisms (56, 197).

PSYCHOPHARMACOLOGY

The accidental discovery in 1943 of the potent psychological effects of LSD, a diethyl derivative of lysergic acid amide, an alkaloid of the ergot fungus (*Claviceps purpurea*) which is also found in the American tropical morning glories (*Argyreia, Convolvulus, Ipomoea, Rivea, Stictocardia* spp.), stimulated a great deal of interest in the biochemical correlates of emotion. Several years later, the less spectacular but more clinically useful effects of reserpine were delineated. Reserpine is an alkaloid of *Rauvolfia serpentina,* a plant which had been used by natives of the Himalayan foothills for centuries to treat a variety of ailments. Since that time, a number of similar, clinically effective so-called "tranquilizing" drugs have been developed—a substantial percentage of the drugs prescribed in the U.S. today fall into this broad category. The therapeutic or curative value of these tranquilizers is limited—their usefulness lies in their ability to create altered emotional states in disturbed patients. This effect is only temporary, and prolonged use of these tranquilizing drugs may result in physical or emotional dependence.

Many other compounds which are functionally similar to LSD have been widely studied, notably the alkaloids mescaline (*Lophophora williamsii* and other species of *Cactaceae*), psilocybin (*Psilocybe, Panaeolus, Conocybe* spp.), harmine (*Peganum, Banisteriopsis* spp.), and synthetics such as methamphetamine, sernyl, and ditran. These compounds have been variously called "psychedelics," "hallucinogens," and "psychotomimetics"— suggesting the deranged, psychotic-like behavior and hallucinosis which they are known to elicit. Unlike the tranquilizers, which are employed to calm the emotional state of the psychotic, no clinical use of these drugs, with the exception of methamphetamine, has been accepted in the United States, and very little in other countries. The hallucinogenic drugs have, however, proven to be effective tools in the investigations of the biochemistry of the central nervous system. These psychopharmacological studies have indicated that emotional states may have certain biochemical correlates, raising the hope of the eventual discovery of methods of altering or treating psychopathological states of mind with true therapeutic agents. Clearly, some pathological mental states are associated with the disruption of central nervous system neurotransmission, as experiments with psychoactive drugs and biochemical research in patients have demonstrated. It is hoped that this line of research will lead to a better understanding of the central nervous system, while developing better techniques to treat mental illness.

Comparison of the chemical structures of various hallucinogenic agents and tranquilizing drugs with the structures of the neurohumors NE, 5-HT, GABA, and glutamate (see Appendix A), suggests close similarities, and it is not surprising that these compounds have a profound influence on the transmission of nerve impulses and, as a consequence, on emotional states. *Generally,* those drugs which reduce the rate or level of stimulation of receptors to neurotransmitters in the brain are associated with depression of emotional mood, while those drugs which cause the rate to increase result in elevation of emotional mood. This distinction, however, is admittedly vague and oversimplified. There are a number of different mechanisms which one could postulate to achieve a *local* increase or reduction in the rate of receptor stimulation. The psychoactive substances in question may enhance or inhibit the synthesis of neurohumors or their release from presynaptic storage vesicles into the synapse. A drug might competitively block receptor sites of postsynaptic neurons, inhibiting the effects of neurohumors, or it may in some manner facilitate the binding of neurohumors, thus potentiating their effects. Other possible mechanisms are the prevention of the reabsorption of diffused neurohumors by the presynaptic neuron, prevention of the uptake of neurohumors by the presynaptic vesicles, or the inhibition of the degradative effects of MAO. Obviously, each of the above mechanisms could effectively result in a localized enhancement or inhibition in the rate of receptor stimulation (15, 21, 36, 85).

Reserpine is an example of a drug which lowers receptor stimulation by neurotransmitters. This drug inhibits the active uptake of NE by the presynaptic storage vesicles, resulting in rapid degradation of NE by MAO. The ensuing deficiency of NE available for synaptic release correlates with a profound depression of emotional mood—this is a classic example of drug-induced depression. Reserpine likewise reduces the brain concentrations of 5-HT and DOPAmine. Chlorpromazine (Thorazine, a widely used tranquilizer) counteracts some psychotic symptoms by blocking receptors to DOPAmine—this action, of course, does not deplete DOPAmine concentrations, but renders ineffective what DOPAmine is present in the synapses exposed to its action. High dietary levels (experimental conditions) of amino acids such as leucine and phenylalanine can induce depression—these amino acids seemingly compete with tyrosine and tryptophan (precursors of NE and 5-HT respectively) for access to permeases in the blood/brain barrier, resulting in inadequate biosynthesis of NE and 5-HT, and eventual depletion of stores. It is supposed that dietary deficiency of tryptophan and tyrosine would have the same effect (15, 36, 85).

The classic model of drug-induced mood elevation is provided by harmine and iproniazid. These compounds inhibit MAO, blocking the degradation of neurotransmitters, which results in an increased concentration of these neurotransmitters in portions of the brain exposed to the action of these drugs. Similarly, imipramine inhibits the reabsorption of diffused NE at the presynaptic membrane, resulting in high concentrations of NE in the synaptic cleft, favoring the accumulation of NE at postsynaptic receptor sites and causing increased stimulation of the postsynaptic neuron—the result is a euphoric state. Cocaine has the same effect as imipramine in the central nervous system. Prolonged use of cocaine and, it is assumed, imipramine, may, however, result in depression—evidently axon biosynthesis of NE, in absence of replenishment by NE reabsorbed from the synaptic cleft, is inadequate to maintain a sufficient supply, and depletion is eventual. Psilocybin and LSD show both receptor stimulation and inhibition of 5-HT, in different parts of the nervous system, and these compounds produce elevation of mood. Both LSD and mescaline inhibit the firing of neurons in the midbrain raphe nuclei, apparently by occupying the binding sites of 5-HT. Mescaline, LSD, and psilocybin, although of disparate chemical structure, show cross tolerance, and produce remarkably similar effects (1, 75).

PHARMACOLOGY OF LSD

The reader is undoubtedly familiar with the psychological effects of LSD and the other hallucinogens, so much has been written about this subject in recent years—this is beyond the scope of this discussion. That these drugs can induce psychotic-like behavior is indisputable; that they are therefore bad is, however, a hotly debated point. In one experiment, psychiatrists were directed to distinguish between tape recordings of "genuine" psychotics and LSD subjects—they were unable to make any such distinction, beyond that expected by chance. In another test, experienced subjects were given doses of LSD and other hallucinogens (doses adjusted to equivalent potency). These users were unable to distinguish one hallucinogen from another, beyond chance. It may be, therefore, assumed, that these compounds are functionally analogous and uniformly produce changes in consciousness. Despite intensive research, the mechanism of LSD action remains obscure. A threshold dose is infinitessimal—30 mcg (30/1,000,000 gram). Immediately after injection the brain contains less LSD than other body tissues—the majority being found in liver, kidney, and small intestine, in the process of being excreted. Within 45 minutes there will be almost no LSD in the brain—by this time, the effects of the drug (which last 8 hours or more) have scarcely begun. In view of the small quantities involved, and the fact that LSD has been effectively eliminated prior to the onset of its effects, it would seem that its role as an antagonist of 5-HT is insignificant. It has been postulated that LSD exerts its effects primarily on the diencephalon and limbic system, areas of the brain which control the autonomic nervous system, regulate emotional awareness, and modulate emotional responsiveness. The phenomenon of visual hallucination may be due to the firing of neurons in the visual reflex area of the brain, in the absence of light stimulus. That this random perception of light which is not really present is often interpreted by the subject as having the appearance of some physical object cannot be explained in biochemical terms. The EEG of the LSD subject shows diminished amplitude, low voltage, complete disappearance of alpha activity. Only 0.01% of an intravenous injection of LSD reaches the brain; 80% is excreted by the liver bile route, 8% is excreted in the urine. The LD_{50} (the dose needed to kill 50% of test animals) in mice is 50–60 mcg/kg, intravenous injection (36, 37, 171).

This is the profile of the best-studied hallucinogen. Obviously, understanding of the mechanisms of hallucinosis artificially induced by chemical agents awaits understanding of the biochemistry and physiology of the central nervous system—this understanding may perhaps come from further study of hallucinogenic drugs.

PHARMACOLOGY OF OTHER HALLUCINOGENIC DRUGS

Harmine, Harmaline (*Peganum, Banisteriopsis* spp.). As indicated above, harmine is an inhibitor of MAO, favoring the accumulation of excess neurotransmitters in synapses of neurons exposed to its action (136). Harmine raised blood pressure and heart rate in sheep and dogs (70). Injections of 5-HT counteracted the effects of harmine, suggesting that its effects are at least partially attributable to antagonism of 5-HT (125). In man, 25–75 mg subcutaneously (below the skin), and 150–200 mg intravenously produced hallucinogenic effects similar to LSD. Oral doses of 300–400 mg, however produced only slight visual symptoms (106, 135).

Atropine/Hyoscyamine, Scopolamine (*Atropa, Datura, Hyoscyamus* spp.). These compounds are cholinergic blocking agents—they competitively inhibit acetylcholine by occupying its binding sites on the postsynaptic membrane, thus preventing excitation by ACh (88, 159).

Lysergic Acid Amide (*Argyreia, Convolvulus, Ipomoea, Rivea, Stictocardia* spp.). This compound is an antagonist of 5-HT, estimated to be, in this regard, 1/25 as potent as LSD in experiments on isolated rat uterus (30). Intramuscular (into the muscle) injection of 5 mg of this drug produced dreamlike euphoria in human experimenters. Oral doses of 2 mg of its isomer, iso-lysergic acid amide, however, produced depression (89).

Macromerine (*Coryphantha macromeris*). This compound is an analog of mescaline, and seems to be considerably less potent. Although experiments on monkeys have shown that this compound is psychoactive, it has yet to be further investigated pharmacologically (87, 115, 116).

bis-**Noryangonin** (*Gymnopilus, Pholiota, Polyporus* spp.). This compound is an analog of yangonin, one of the psychoactive agents of the South Pacific "kava kava" (*Piper methysticum*), and has been isolated from fungi of these genera. Some of these fungi are known to have psychoactive properties. This drug itself, however, has not been tested pharmacologically (78).

Tetrahydrocannabinols (*Cannabis* spp.). Very little is known about this unique class of hallucinogens (they are non-nitrogenous) on the neurochemical level. Physiologically, THC produces acceleration of heart rate, increased blood pressure, vertigo, increased sensitivity to touch, decreased body temperature, and increase in appetite (159). Tests have shown that 1.5 hours after intraperitoneal (into the abdomen) injection in rats, most of the THC was in the liver (122). In rats, whole brain 5-HT increased considerably after injection of THC (16). THC retarded depletion of 5-HT by reserpine, but showed no MAO interactions or effects on the NE system (69, 109, 110).

Mescaline (*Lophophora, Trichocereus* spp.). Although this compound shows cross tolerance with LSD, it has been shown to potentiate the effects of 5-HT in frog uterine muscle (38). It shows an inhibitory effect on serotonergic neurons in the central nervous system (1, 40). Doses of 200–600 mg produce effects almost identical to LSD in human subjects, with a side effect of nausea (53). This compound is readily absorbed from the intestinal tract, and causes increase in body temperature (107, 137). The

LD$_{50}$ in mice is 500–600 mg/kg, death from respiratory failure (53). In man, 60–90% is excreted unaltered in the urine, the remainder is excreted as 3,4,5-trimethoxyphenylacetic acid (66).

Psilocybin, Psilocin (*Psilocybe, Panaeolus, Conocybe* spp.). These compounds are antagonists of 5-HT, and show cross tolerance with LSD (29, 96). Dosage of 4–8 mg of psilocybin (the equivalent of approximately 2 g dry weight of *Psilocybe mexicana*) produces effects similar to LSD in man. Psilocin is approximately 1.4 times as potent as psilocybin, suggesting that dephosphorylation (see Appendix A) must occur before this compound can enter the brain (79). In fact, injections of psilocybin resulted in the accumulation of psilocin in kidney, liver, and brain. Behavioral effects closely paralleled the increase in brain concentration of psilocin (93). Approximately 50% of labeled psilocin was absorbed by the gastrointestinal tract; distribution in body tissues was uniform after 30 minutes (79). Both compounds cause increase in body temperature, dilation of pupils, erection of hairs, and contraction of nictitating membrane (29). The LD$_{50}$ in mice is 280 mg/kg (90). Injections of psilocybin result in the excretion of psilocin—65% in urine, 20% in bile and feces after 24 hours. After 7 days, psilocin could still be detected in the urine (93, 99).

Asarone, beta-Asarone (*Acorus calamus*). These stereoisomers have been isolated from a plant which is known to be psychoactive, producing analgesia in low doses and hallucinosis in higher doses. Asarone has been shown to be psychoactive, but may not be the agent in the plant responsible for hallucinosis. The evidence seems to indicate that this compound has rather a depressive effect. Asarone potentiates the effects of reserpine and chlorpromazine, not, however, by influencing 5-HT concentrations (41). No analgesic activity has been demonstrated for asarone. Both isomers prolong the anesthetic effects of ethanol, pentobarbitone, and hexobarbitone (161). Asarone may be anti-cholinergic in the peripheral nervous system. Both isomers caused decreased body temperature, cardiac depression, decreased blood pressure. Near toxic doses caused ataxia, hypnosis, loss of righting reflex—beta-asarone showed none of these effects at any dose (42).

Ibotenic Acid, Muscimol (*Amanita* spp.). Much confusion surrounds these mushrooms, their use and toxicity. Both of these compounds have been isolated from hallucinogenic *Amanita* spp., and shown to be psychoactive (183). Muscimol perhaps does not exist in the live mushroom

−it is a decarboxylation product of ibotenic acid (see Appendix A), and this decarboxylation readily occurs during any attempt at chemical manipulation. As is the case with psilocybin/psilocin, the breakdown product is stronger, muscimol being 5–10 times as potent as ibotenic acid, suggesting that only muscimol reaches the brain (183). Both compounds influence brain concentrations of NE, 5-HT, and DOPAmine, in a manner similar to LSD (44, 101). Muscimol is an analog of GABA (see Appendix A) and, like GABA, is a powerful inhibitor of central nervous system neurotransmission (98). Ibotenic acid is an analog of glutamate (see Appendix A), and seems to mimic its effects. Ingestion of muscimol, by human subjects, in doses of 7.5–20 mg caused hallucinosis, often combined with muscle spasms and nausea (175). I have ingested *A. muscaria* and *A. pantherina*, the latter producing strong hallucinosis characterized by slight visual distortions (130). Both ibotenic acid and muscimol have been recovered from the urine of human subjects, potential verification of the reports of Siberian travelers (32). My experiments on mice showed that only 5–10% of intraperitoneally injected muscimol could be recovered in the urine within 48 hours. Several breakdown products were detected (131). Further tests on human subjects showed that a substantial percentage of orally ingested ibotenic acid could be recovered in the urine within 90 minutes (32). In the Siberian practice (see Part I), some of the excreted ibotenic acid was evidently decarboxylated to muscimol in the bladder, or, subsequently, in the gastrointestinal tract of the drinker of the urine. As muscimol is 5–10 times the potency of ibotenic acid, only a fraction need decarboxylate to muscimol during the ingestion of urine of a mushroom eater−the remainder being re-excreted. It has been reported that in Siberia, one dose of mushrooms could be recycled through 4 or 5 persons (189). Atropine potentiates the effects of these toxins−it is widely used as an antidote to accidental mushroom poisonings (as many of these poisonings involve mushrooms containing high concentrations of muscarine, for which atropine is an effective antidote), often with unfortunate consequences, in the case of *Amanita* poisoning. The toxins seem to be concentrated in the skins on the caps of *Amanita* mushrooms. Those who use these mushrooms as recreational drugs often peel these skins and smoke them dried. Further, those persons who eat hallucinogenic *Amanita* species as food are known to peel the skins off and discard them, prior to parboiling the remainder of the mushrooms and discarding the water−this treatment renders the mushrooms non-toxic, as the toxins are water soluble. There seems to be some relation between the pigmentation of the cap and toxicity (130).

Nicotine (*Nicotiana* spp.). Although the biochemical effects of nicotine on the autonomic system have been well studied, its central effects have only recently come under scrutiny. Nicotine has been shown to affect the levels of 5-HT in the brain (124). Further, nicotine has been shown to have a profound effect on the metabolism of niacin, or nicotinic acid, a related compound which is a B group water soluble vitamin (anti-pellegra factor) (59). Dietary deficiency of niacin results in psychosis, as well as other symptoms. Massive doses of niacin and niacin amide have been used with some success in the treatment of schizophrenia. Harman and norharman, derivatives of the indole hallucinogen harmine, have been isolated from tobacco leaves, and in higher concentrations in tobacco smoke (138). These compounds have not been tested pharmacologically. Clearly, there is some evidence to support, on a chemical basis, the contention that plants of the genus *Nicotiana* are hallucinogenic. Plants of this genus are still used as ritual hallucinogens in South America, and have a long history of ritual use throughout the New World (97, 198). Why then, is the use, in our culture, of tobacco not associated with hallucinosis or altered states of consciousness? Tobacco is used with such frequency by "civilized" smokers that, due to the possibility that tolerance to its hallucinogenic effects may be quickly acquired (as is the case with many of the other hallucinogens), it does not produce alterations of consciousness or hallu-cinations. Further, it is apparent that hallucinosis and consciousness al-teration are very much a function of the expectation of the user; that is, it happens that users who have no expectation of "getting high," who do not in fact know what it means to feel "high," may use hallucinogenic drugs and feel absolutely no effect (194). This phenomenon is commonly observed with marijuana—it has also been observed with psilocybin, LSD, and other drugs. My recent study on accidental and intentional ingestion of hallucinogenic *Amanita* species has further demonstrated this pheno-menon (132). Indeed, altered states of consciousness virtually identical to those induced by hallucinogenic drugs may be produced through the agency of hypnosis, biofeedback training, or meditation—suggesting that altered states of consciousness are not pharmacological effects at all, though they are often associated with the use of certain drugs, which apparently serve as keys or triggers to these alterations in consciousness.

NOTES ON HALLUCINOSIS BEYOND THE DRUG STATE

In recent years, as the use of hallucinogenic drugs has become more widespread, several writers, notably Timothy Leary, have pointed out the close correlation between the "psychedelic experience" and the religious or ecstatic states attributed to the practice of Oriental religions, such as Buddhism, Hinduism, and Zen. This is, of course, more than a coincidence. Leary went so far as to edit a version of the *Bardo thödol,* the so-called *Tibetan Book of the Dead,* pointing out this correlation, with the implication that LSD and the other hallucinogens were short cuts to religious ecstasy. The appeal of this argument was to a higher authority, as if there were, objectively, a religious state which hallucinogens could mimic; when in fact these ostensibly spiritual states, the folklore surrounding them, the very foundation of religious ideas all evidently stem from the consciousness alterations induced by the accidental ingestion of hallucinogenic plants by primitive men. The visions being described in the holy books of today's religions probably are hallucinogenic visions, produced by magic plants; the practices associated with Oriental religions may be simply attempts to produce consciousness alteration without the use of plants, though, long centuries having passed since the plants were last used, this may of course be unknown to the participants in the pageantry of today's religions. A classic example is the Soma cult, described in the *Ṛg Veda.* This cult came into being some 4000 years ago. While the rituals and hymns involving the use of the Soma plant still exist, the use of the plant itself has long since ceased, and various substitute plants, none of which are hallucinogenic, have been used over the years (189). The various techniques which have come to be called yoga—fasting, sensory deprivation, meditation, breath control—have grown out of the religion which originated in the Soma cult, and have doubtless come into being in an attempt to produce alteration of consciousness without the use of psychotropic plants (see Part Two). This attempt was perhaps stimulated by the scarcity of the plants, vagaries of seasonal supply, or domination of the use of plants by a theocracy. Although the reason is not known, it is evident that the use of Soma ceased in India, and a religion characterized by highly evolved techniques for producing consciousness alterations grew out of the remains of the Soma cults.

Transcending the necessity of using plants to alter consciousness was accompanied by transcending the limitations of their use—the inevitable cessation of consciousness alteration, when the drug responsible for this was excreted. It is apparent that the alterations of consciousness produced by yoga practices are much more profound and meaningful than those produced by the ingestion of psychotropic plants—drug users typically become interested in yoga, may become practitioners of yoga techniques, and may even cease using drugs. The reverse never occurs; that is, those who can alter consciousness without the use of drugs have no interest whatever in their use. At the present time, it is possible to postulate biochemical mechanisms for the induction of hallucinosis by fasting, and some recent research has been conducted on the mechanisms of hallucinosis produced by hypnosis, sensory deprivation, and hyperventilation.

It has been demonstrated that sensory deprivation produces effects which are remarkably similar to the prototype LSD state. Interesting interactions between the two states have been observed (37, 108, 205). Sleep deprivation, similarly, produces hallucinosis. The induction of hallucinosis by hypnosis is a well known phenomenon. It is also possible for drug users to relive past experiences with hallucinogenic drugs by using hypnotic age regression techniques (4). Recently the physiological effects of hyperventilation have been studied, in particular the induction of hallucinosis through the use of hyperventilation (6). As mentioned earlier, control of breathing, or pranayama, is an integral part of the yoga techniques used to produce alterations in consciousness. None of the above techniques are well understood at this time; clearly, further research is needed to determine how these methods can produce hallucinogenic states.

It has long been apparent that nutritional pathology results in altered states of consciousness. Yogis have used fasting for centuries to produce consciousness alteration. Most of the known nutritional pathologies which affect the brain have a direct influence on the brain's metabolism of energy-yielding dietary substances. In complete fasting, the brain, which depends exclusively on glucose for its energy supply under optimal conditions, must adjust to an energy supply which is below normal and consists of metabolites other than strictly glucose. The thiamine (Vitamin B_1) deficiency disease, beriberi, is characterized by disruption of energy metabolism throughout the body. Thiamine, as thiamine pyrophosphate, is a coenzyme of carbohydrate metabolism. That is, this vitamin is necessary to process dietary sugars into molecules which the cells of the body can use for energy. Thiamine pyrophosphate is vital to at least three steps in the TCA cycle (tricarboxylic acid cycle—energy yielding cycle which causes the "burning" or oxidation of simple sugars to yield carbon dioxide and ATP, the "fuel" which drives biochemical reactions). Disruption of the TCA cycle by deficiency of thiamine will result in psychotic symptoms—long recognized as characteristic of beriberi. Psychotic manifestations, similarly, are a symptom of pellagra, which is caused by a deficiency of niacin. This vitamin, as nicotinamide-adenine-dinucleotide, is another coenzyme vitally important in carbohydrate metabolism. At lease four steps in glycolysis (breakdown of dietary carbohydrates into metabolites for the TCA cycle) and the TCA cycle are dependent on this coenzyme. Pyridoxine, as pyridoxal phosphate, is also vital to glycolysis and the TCA cycle—a deficiency of pyridoxine likewise results in psychotic states. Cobalamin, Vitamin B_{12}, is evidently vital to the brain as well—deficiency of this vitamin, called pernicious anemia, is characterized by neurological lesions. Deficiency of pantothenic acid also results in brain pathology and psychosis. This vitamin is a component of Coenzyme A, which is necessary for the oxidation of the products of glycolysis in the TCA cycle. Prolonged deficiency of the minerals sodium, potassium, and magnesium will result in nervous system pathology. The reader will recall that it is the maintenance of proper ionic balances of sodium and potassium, and the flux of these ions across the axon membrane which is responsible for the conduction of nerve impulses by the axon. It is not surprising that deficiencies of these minerals can cause paralysis of the extremities and, potentially, psychotic manifestations. Magnesium ions are necessary for the active uptake of NE by presynaptic storage vesicles. Deficiency of this mineral is known to produce nervous system pathology, perhaps by producing a depletion of storage pools of NE available for release into the synapse (105, 193).

It must be stressed that the above mechanisms are hypothetical. There must necessarily be a whole complex of symptoms attending the deficiencies of the above-named nutrients; the mechanisms which have been postulated are, however, plausible. There is no doubt that these deficiencies result in neuropathological symptoms.

CONCLUSION

It is obvious that emotion, thought, memory, and similar concepts are not understood on the biochemical level. Questions of altered emotional states are necessarily obscure. At this stage there are mysteries of the higher nervous system which elude physical understanding. It may be that man's brain, for lack of an outside frame of reference, may never be able to understand itself fully.

APPENDIX A

CHEMISTRY OF HALLUCINOGENS AND NEUROTRANSMITTERS

A little poison now and then: that makes for agreeable dreams.
And much poison in the end, for an agreeable death.

—FRIEDRICH NIETZSCHE, *Thus Spake Zarathustra*

Chemical structures enclosed in boxes are known and putative neuro-transmitters (see Part III). These are drawn in such a manner, and juxtaposed with the structures of various hallucinogens to show structural similarity.

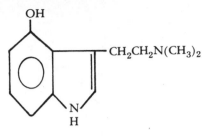

PSILOCYBIN
4-phosphoryloxy-dimethyltryptamine
(*Psilocybe, Panaeolus, Conocybe* spp.)

PSILOCIN
4-hydroxy-dimethyltryptamine
(*Psilocybe, Panaeolus, Conocybe* spp.)

SEROTONIN
5-hydroxy-tryptamine

LAA
LYSERGIC ACID AMIDE
(*Argyreia, Stictocardia, Rivea,
Ipomoea, Convolvulus* spp.)

LSD
LYSERGIC ACID DIETHYLAMIDE
(Synthetic)

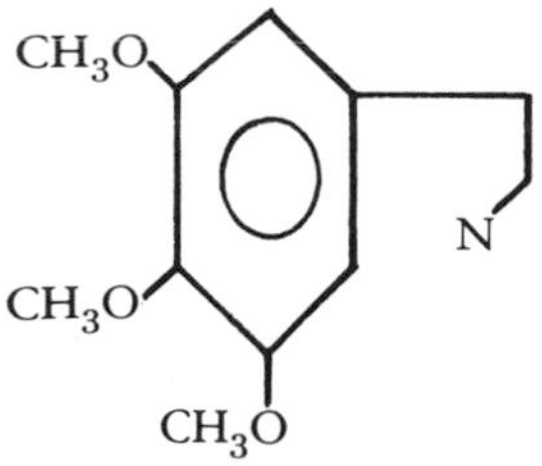

MESCALINE
3,4,5-trimethoxy-phenethylamine
(*Lophophora, Trichocereus* spp.)

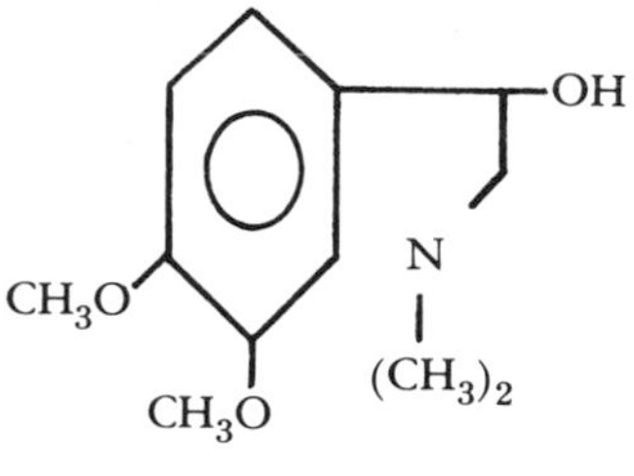

MACROMERINE
α-[(dimethylamino) methyl] veratryl alcohol
(*Coryphantha* spp.)

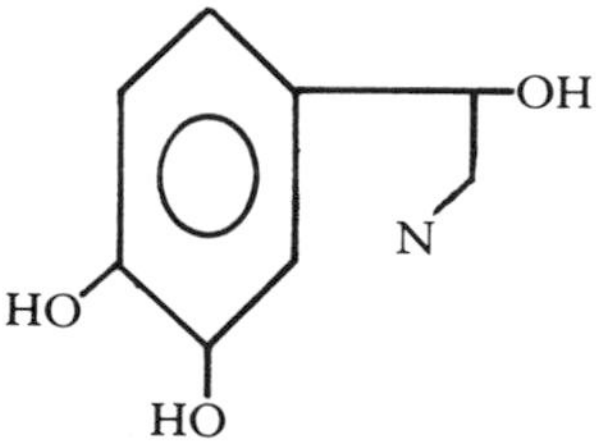

NOREPINEPHRINE
α-(aminomethyl)-3,4-dihydroxybenzyl alcohol

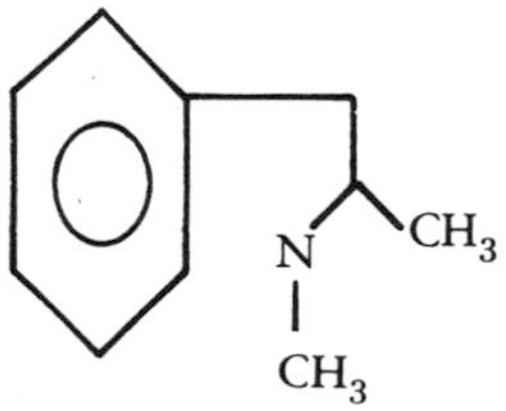

METHAMPHETAMINE
d,l-α-methylphenethylamine
(Synthetic)

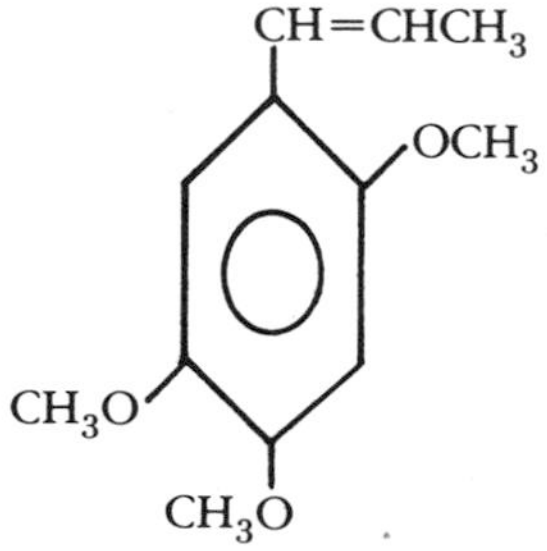

ASARONE
2,4,5-trimethoxy-1-propenyl benzene
(*Acorus calamus*)

IBOTENIC ACID
α-amino-3-hydroxy-5-isoxazole acetic acid
(*Amanita* spp.)

MUSCIMOL
3-hydroxy-5-aminomethyl isoxazole
(*Amanita* spp.)

GLUTAMATE
2-aminopentanedioic acid

GABA
4-aminobutanoic acid

BUFOTENINE
5-hydroxy-dimethyltryptamine
(*Amanita, Bufo, Anadenanthera* spp.)

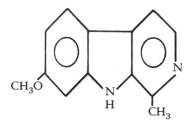

HARMINE
7-methoxy-1-methyl-9H-pyrido[3,4-b] indole
(*Peganum, Banisteriopsis* spp.)

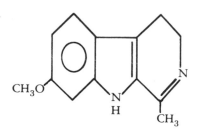

HARMALINE
4,9-dihydro-7-methoxy-1-methyl-3H-pyrido[3,4-b] indole
(*Peganum, Banisteriopsis* spp.)

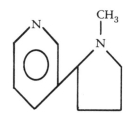

NICOTINE
1-methyl-2-(3-pyridyl) pyrrolidine
(*Nicotiana* spp.)

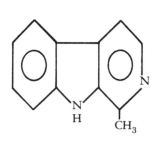

HARMAN
1-methyl-9H-pyrido[3,4-b] indole
(*Nicotiana* spp.)

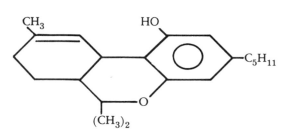

THC
Δ^1-3,4-*trans*-tetrahydro cannabinol
(*Cannabis* spp.)

bis-**NORYANGONIN**
4-hydroxystyryl-2-hydroxy-4-pyrone
(*Gymnopilus, Pholiota, Polyporus* spp.)

126

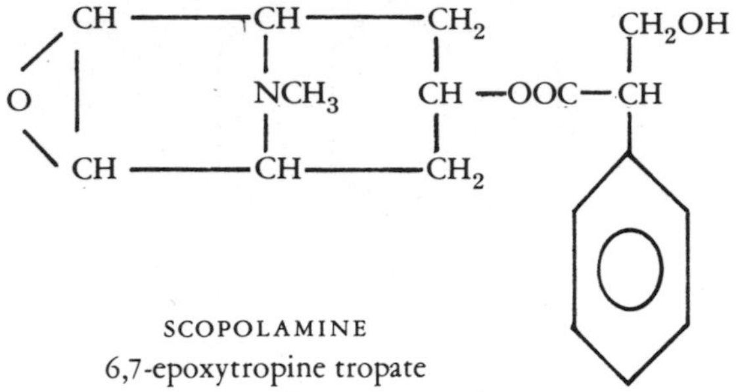

$$CH-CH-CH_2$$
$$O \quad | \quad NCH_3 \quad CH-OOC-CH \quad CH_2OH$$
$$CH-CH-CH_2$$

SCOPOLAMINE
6,7-epoxytropine tropate
(*Atropa, Datura, Hyoscyamus* spp.)

$$CH_2-CH-CH_2$$
$$| \quad NCH_3 \quad CH-OOC-CH \quad CH_2OH$$
$$CH_2-CH-CH_2$$

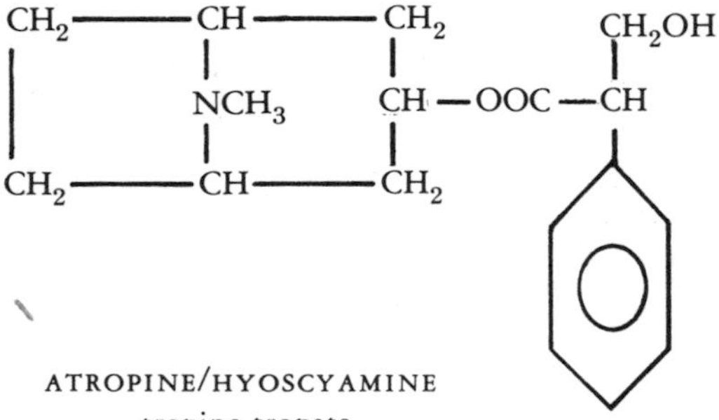

ATROPINE/HYOSCYAMINE
tropine tropate
(*Atropa, Datura, Hyoscyamus* spp.)

$$CH_2-CH-CHCOOCH_3$$
$$| \quad NCH_3 \quad CHOOC$$
$$CH_2-CH-CH_2$$

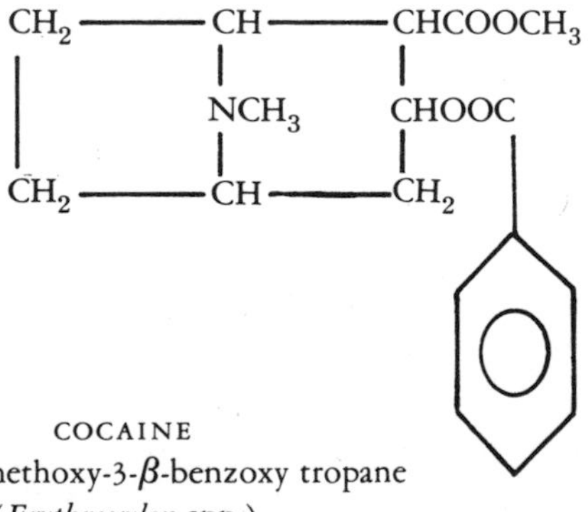

COCAINE
2-β-carbomethoxy-3-β-benzoxy tropane
(*Erythroxylon* spp.)

CHLORPROMAZINE
2-chloro-10-(3-dimethylamino propyl) phenothiazine
(Synthetic)

RESERPINE
3,4,5-trimethoxybenzoyl methyl reserpate
(*Rauvolfia* spp.)

APPENDIX B

WORDS

To see a World in a Grain of Sand
And a Heaven in a Wild Flower,
Hold Infinity in the palm of your hand
And Eternity in an hour.

—WILLIAM BLAKE, "Auguries of Innocence"

ABSCISSION LAYER: separation layer

ACUMINATE: with a sharp pointed leaf tip tapering concavely to the leaf body

ACUTE: with a sharp pointed leaf tip tapering straight or convexly to the leaf body

ACHENE: a small, dry, indehiscent, one-seeded fruit

ADNATE: gills of a mushroom adhering to the stipe

ADNEXED: gills of a mushroom reaching the stipe, but barely connected

ALKALOID: an organic substance having alkaline properties and containing nitrogen; a biodynamic compound isolated from a plant

ALLOSTERIC EFFECT: the communication of a conformational change from one protein subunit to the next, as in a membrane

ANNULUS: a membranous or fleshy ring which surrounds the stipe of a mushroom after the expansion of the pileus; remnants of the partial veil adhering to the stipe

ANTHERS: the parts of a stamen that contain pollen

APEX: tip or end of a part

ATP: adenosine triphosphate, a molecule having high-energy phosphate bonds, which mediates cellular energetics

AXIS: the main stem

AXILLARY: solitary

AXOPLASM: the cell contents of a neuron bounded by the axon membrane

BACCATE: berry-like

CALCAREOUS: containing lime, calcium carbonate, or calcium

CALYX: all the sepals together in a flower

CAMPANULATE: bell-shaped

CARPELLARY: having carpels, pistils that are leaf-like

CARPOPHORE: the fruiting body of a fungus, a mushroom

CHROMATOGRAPHY: chemical separation achieved by dipping a sheet onto which a mixture of compounds has been spotted, into a solvent, and allowing the solvent to wick up the paper

CIRCUMSCISSILE: opening by transverse fissure around circumference (as a seed) leaving upper and lower half

COMATE: covered with hair or filaments

CORDATE: heart-shaped leaf

COROLLA: all the petals together in a flower

CYME: a cluster of flowers

CYMOSE: with a convex cluster of flowers

DECIDUOUS: shedding leaves annually

DECUMBENT: trailing on the ground and raising at the tip

DECURRENT: extending down along the stem

DENTATE: with sharp indentations or teeth perpendicular to the margin of a leaf

DEPOLARIZATION: destroying or counteracting a state of polarization, or charge distribution

DEPRESSED: flattened

DIGITATE: fingerlike or handlike; leaves sprouting from one point

ENTIRE: with no indentations around the leaf margin

FLOCCULENCE: tufts of soft wooly hairs

GLABROUS: not hairy

GLAUCOUS: covered with a whitish substance which rubs off

INDEHISCENT: staying closed

INFERIOR: below, underneath

ISOMERS: chemical compounds having the same elemental composition, but different structures

LANCEOLATE: lance-head shape

LIANA: woody, tropical climbing vine

LIMB: expanded section of a leaf or petal

LINEAR: long, narrow, and flattened with parallel edges, like grass blades

MYCELIUM: hyphae of a fungus; cobweb-like threads of a mushroom which pervade its substrate or nutrient supply

MYCORRHIZAL: a symbiotic (or parasitic) association between a mycelium of a fungus and the rhizome or root of a higher plant

OVATE: egg-shaped

PEDUNCLE: a leafless axis bearing several flowers

PENDULOUS: drooping

PENTAFID: divided into five parts

PERFECT: bisexual flower

PERIANTH: calyx and corolla together

PETALS: leaf-like structure composing the corolla

PETIOLES: leaf stalk

PILEUS: the cap of a mushroom

PINNATIFID: having leaves in a feather-like arrangement, with narrow lobes whose clefts extend more than half-way to the stem

PISTIL: seed-bearing organ of a flower; ovary, stigma, and style

POLYCEPHALOUS: having many heads

PUBESCENT: hairy, hirsute

PUBERULENT: covered with exceptionally short hair perpendicular to the surface and not easily visible to the eye

RHIZOME: underground creeping stem; root

SCARIOUS: thin and dry

SCAPE: a leafless axis bearing a flower

SEPAL: green leaf-like structure composing the calyx

SERRATE: with sharp, forward-pointing teeth

SESSILE: stalkless

SETACEOUS: having bristles

SINUATE: gills of a mushroom attaching to the stem, but leaving an indentation or sinus at the point of attachment

SPADIX: a single axis bearing a stalkless flower

SPATH: leaf-like structure enveloping the spadix

SPORE PRINT: made by cutting off the pileus of a mushroom and placing with gill surfaces down on clean white paper; cover with a glass and allow time for spores to drop onto paper

STAMEN: pollen-bearing organ of a flower, having filament and anthers

STIPE: the stem of a mushroom

STIPULES: basal appendages of the leaf stalk

STYLE: stalk-like part of a pistil between stigma and ovary

SUBEQUAL: approximately but not exactly equal

SUBGLOBOSE: imperfectly or nearly globose

SUCCULENT: fleshy or juicy

SUPERIOR: above or upper surface

TERATOGENIC: producing birth effects

TOMENTOSE: covered with densely matted hairs

TUBERCULES: bulbous, knob-like growths of certain plants, especially cacti

UMBO: a rounded or pointed elevation in the center of the cap of a mushroom

UNIVERSAL VEIL: a membrane that initially covers the young sporophore of various mushrooms and is ruptured by growth—in mature mushrooms may persist as a volva, or as warts or scales on the surface of the cap

VISCID: covered with a glutinous or sticky layer

VOLVA: a membranous bulb that surrounds the base of the stem of certain mushrooms and is formed by the rupture of the veil

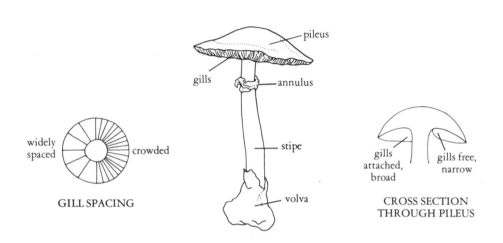

GILL SPACING

CROSS SECTION THROUGH PILEUS

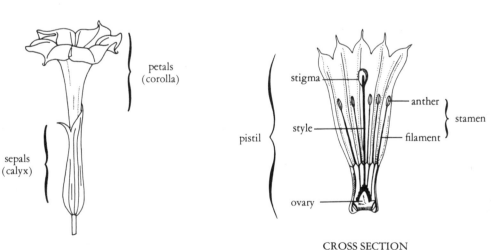

CROSS SECTION THROUGH FLOWER

APPENDIX C

FIELD AND EXTRACTION TECHNIQUES
FOR THE AMATEUR

The ten fingers, the two arms, harness the pressing stone; they are the preparers of the Soma, with active hands. The one with good hands has milked the mountain-grown sap of the sweet honey; the aṃśú has yielded the dazzling.

—*Ṛg Veda*, Maṇḍala V, 43[4]

FIELD TECHNIQUES

VASCULAR PLANTS

It is desirable to construct a plant press, to aid in the collection, identification, and preservation of plant samples. The press is simply a stack of pieces of cardboard (corrugated is best) with folded sheets of newspaper in between the sheets of cardboard. Top and bottom covers should be made of wood, with holes drilled to provide ample ventilation. The covers, cardboard, and newspaper are held together by two leather straps or lengths of rope, which can be secured around the press to provide tension. The straps or rope should be long enough to allow for expansion of the press when it is full of plants. A branch of the plant in question, with leaves and flower (if possible) should be placed in between the folded sheet of newspaper between two sheets of cardboard and kept in the tightened press until dry. It may be necessary to replace the newspaper as it becomes wet. Of course, during drying, the press should be placed in a warm, dry place. This technique will result in the preparation of a preserved sample of the plant, for future reference.

Xerox the accompanying sheet on field techniques and note all of the details for each plant, assigning the specimen a number which is recorded on the worksheet. Comparing this information, as well as the specimen with the plant descriptions in the text will result in easy identification. When you are sure of the identification, collect those portions of the plant which you wish to use. The press, as well as paper or plastic bags for collection, a notebook with worksheets and a pen, and a small ruler with metric calibration, can all be easily carried along.

It is advisable to carry small wax or paper bags, to isolate individual types of mushrooms, and a large wicker basket, to carry these bags. Do not use plastic bags, as they do not "breathe," and will cause the mushrooms to decompose rapidly. Do not put more than one species in any given paper bag.

Again, carry Xerox copies of the worksheet, and note all pertinent information on each species collected, placing the information so recorded for each species in the bag with the mushrooms. When you return from the collecting trip, lay all of the mushrooms out to dry on newspaper, and prepare a spore print for each species, in the following manner: carefully sever the cap from the stem and place it, gill surface down, on a clean portion of the worksheet describing that particular specimen. Cover the inverted cap with a clean glass or bowl, and allow it to remain undisturbed for several hours, perhaps overnight. Gravity will cause the spores to drop onto the paper, forming what is called a spore print. Note the color of this print, and enter this information in the appropriate blank on the worksheet. It is advisable to exercise extreme caution in identification, until you are certain that you know a given species. It is also advisable to become familiar with the few species of potentially lethal mushrooms which may grow in your area. *Galerina* species could conceivably be mistaken for *Psilocybe* by the untrained and incautious collector. *Galerina autumnalis* and other species of *Galerina* contain alpha-amanitin, which is deadly toxic. There is a photograph of *G. autumnalis* included in the plate section, as well as a brief description of the identifying features (see PLATE 13). *Galerina* species have rust-brown spores, while the *Psilocybe* species have purple-brown spores. There are also mushrooms in the genus *Amanita* which contain alpha-amanitin, and which are potentially lethal, notably *A. phalloides*. See PLATE 3 for a photograph of this mushroom, as well as a description of the features which distinguish this mushroom from the hallucinogenic *Amanita* species.

EXTRACTION TECHNIQUES

Place the fresh or dried plant material in a blender, and mix with methyl alcohol (methanol) for several minutes. It may be desirable to cut or crush the material into small pieces to aid extraction. Pour the resulting extract through a "Melitta" coffee filter, or other similar paper filter, and discard the material retained by the filter.

Pour the resulting solution into a pan with a large, flat bottom. Allow to stand open overnight, or until the alcohol evaporates. When the material is dry, redissolve in a small amount of water, and evaporate again by the same procedure. This is to drive off all of the methanol—*it is highly poisonous.* This second evaporation should be done if you intend to eat the resulting extract. NOTE:*Keep methanol and all extraction solutions away from open flames, as these are highly flammable.*

WORKSHEET FOR FIELD IDENTIFICATION

Sample Number: *Date:*   *Weather Conditions:*

VASCULAR PLANTS:
1. General size, shape, color, description:

2. Habitat and location:

3. Leaves: *Color:*   *Shape:*   *Size:*

4. Flower: *Color:*   *Shape:*   *Size:*

    Petals (color and number):   *Number of stamens:*

    Number of sepals: *Length of stamens:*

    Length of style:

5. Fruit: *Color*   *Shape:*   *Size:*

6. Seeds: *Color:*   *Size:*   *Number:*

MUSHROOMS:
1. Habitat and location:

2. Nutrient supply (i.e. dung, wood):

3. Pileus: *Size:*   *Color of surface:*   *Color of flesh:*

    Shape: *Presence of umbo:*   *Presence of warts:*

4. Stipe: *Size:*   *Color of surface:*   *Color of flesh:*

    Presence of annulus: *Color of annulus:*

    Presence of volva:

5. Gills: *Color:*   *Attachment to stipe:*   *Narrow or broad:*

    Crowded or spaced:

6. Spores: *Color:*

7. Odor:

8. Taste:

140

AFTERWORD

The universe is wider than our views of it.

—Henry Thoreau, *Walden*

This book represents an attempt at a scientific study of hallucinogenic plants. There are other ways to study these plants. Science, perhaps, offers us the only systematic way, however, it can not, at this time, enable us to know how these plants produce changes in consciousness. A man coming to these pages with no experience of the altered states of consciousness produced by psychotropic plants may or may not find this material interesting. He can in no way expect to understand it, that is, to derive meaning from it—apart from a verbal, superficial comprehension.

Yet, quite clearly, man has a long history of deriving meaning from these plants; indeed, the religious institutions of our history, thought by some to represent the zenith of our accomplishments as animals, those very institutions for which men have lived and died, in whose names vast empires have been subjugated, wars fought, whole races enslaved, owe their existence to the primordial ingestion of hallucinogenic plants by aboriginal man. Religions and wars come and go, mere ripples on the face of the past, yet this meaning remains and is quite as accessible to us today as it was, wherever, whenever, to the priests of Xochipilli; of Soma.

Based on the altered perceptions of men who have used these plants, whole systems of belief have been constructed which, however alien to scientific reason, however "absurd" they may seem in the light of our scientific sobriety, are nonetheless complete systems. That is, those that manipulate these systems find that their inner "logic" has predictive value, and may be applied successfully to such concrete problems as healing the sick and solving crimes. Surely, it is easy to dismiss these "separate realities" (Castaneda) as mere "smoke of opinion" (Thoreau), as superstitious nonsense; it will be a much more difficult task to learn what we can from them, while there is still time. It has been suggested that ritual healing, accepting as the etiology of illness sundry mechanisms quite alien to the cherished "germ theory," depends, for whatever curative value it may possess, on the faith of the patient in its efficacy; this faith is regarded, with condescension, as naive gullibility. We commonly forget that modern medicine, too, depends on the patient's faith in the system being applied, and no type of therapy can aid a person who believes his illness is insurmountable; who believes he will die, or wants to.

Despite the fact that the world, as the user of hallucinogens sees it, is vastly different from the world most of us ordinarily see, *it is no less real.* What we take to be reality is the world as we have been taught to see it – it is, by a process of integration and interpretation of sensory input, a construction of the sensory apparatus, and no less an hallucination than the world as we may see it under the influence of hallucinogens. We consider the "normal" perception of the world to be "reality" because our common conditioning places us in consensus with the majority of our fellows, and those persons who otherwise perceive the world are taken to be "mad" or "sick." We should be very hesitant to dismiss as false symbol systems derived from an, in some way, altered perception of the world – simply because their "logic" is alien, apart from our culturally conditioned consensus. It is evident that those who understand the world through an altered state of consciousness may be, among themselves, in consensus regarding their perceptions. Our integration of sensory input is much like a language we are taught – as long as two people have been taught the same language, they can communicate through it.

As a reader of this book, you are obviously interested in hallucinogens. This book may have taught you some facts, it may have interested you in further study of hallucinogens, but it cannot have brought any real meaning into your life unless you know the experience that binds this material together, which binds us together as writer and reader – the hallucinogenic experience. This book is, at best, but a topographical description of a world of experience, presented in a coordinate system which will enable those schooled in this system to know the features of this world – there is, however, a difference between *knowing* an experience and *having* it.

Constructive experiences with hallucinogens are worth scores of books like this. They may represent for us fresh breaths in our stuffy lives, they may produce revelations, they may, at least, enlighten us as to the nature of our perceptions; the absolute "sanity of common sense" (Laing). My task has been to document studies relating to these experiences. Knowing that certain of my readers do not share with me this realm of experience, I feel, in retrospect, as though I have been describing the sun, seen through stained glass windows, to a man born sightless.

Sure, I talk about the glow, the softness, texture, warmth . . . yes, even detail the respective frequencies of light and their intensities and spatial array . . . but the man just doesn't *see* it, you know . . . and if I try too hard, he might think I'm crazy!

Jonathan Ott
Seattle-Olympia-Mexico City
January 1974–January 1976

ACKNOWLEDGEMENTS

As would be natural for any scholar engaged in multi-disciplinary research, I am much in debt to specialists in various fields, whose advice and assistance has enabled me to bring this work to fruition. It is my pleasure to here express my gratitude to Dr. William Scott Chilton of the University of Washington, for guiding my research in both laboratory and library, for reviewing and criticising my manuscript and for helping me through the many lean months of my undergraduate career.

I am most grateful to Mr. R. Gordon Wasson of Danbury, Connecticut for taking such a kind interest in my work, for providing me with photographs and rare reference sources, and for reviewing and criticising my writing. Mr. Wasson's tireless and remarkable work in ethnomycology has been a constant source of inspiration to me.

Jeremy Bigwood of the Fitz Hugh Ludlow Memorial Library has worked energetically on behalf of this book. His broad knowledge of the history of hallucinogenic plants has been a valuable resource. I am indebted to him for his criticism of my manuscript, his assistance in locating reference material, and his excellent photographic work.

I am grateful to Dr. Richard Evans Schultes of the Botanical Museum of Harvard University for reviewing the manuscript and for lending his name to my work in its early stages—this opened many doors for me.

I also wish to express my gratitude to Dr. Steven H. Pollock of the University of Texas Health Science Center at San Antonio for valuable advice and assistance, and to Paul Vergeer of the San Francisco Mycological Society, for reviewing my manuscript and suggesting corrections.

Dr. Gastón Guzmán of the National Polytechnic Institute of Mexico has always been ready to assist me in mycological matters. I also thank Dr. Daniel E. Stuntz of the University of Washington and Marcella Pearsall of Olympia, Washington for mycological advice.

I am indebted to Dr. Lynn R. Brady of the University of Washington, Dr. Andrew Weil of the Botanical Museum of Harvard University, Dr.

Michael Aldrich of the Fitz Hugh Ludlow Memorial Library, John Bailin of the National University of Mexico, and Elliot Marks of the Washington Governor's Office for technical assistance, and to Dr. Elizabeth Kutter, Dr. Linda Kahan, and Dr. Burton S. Guttman of The Evergreen State College for academic sponsorship of the project.

I wish to thank Joy Spurr and Ben Woo of the Puget Sound Mycological Society, Jerry Boydston, Robert Gerrish and Barry Roderick of The Evergreen State College, Dr. J. L. McLaughlin of Purdue University and Robert Harris of Inverness, California for supplying photographs of the plants.

For a fellowship that made possible my research in Mexico I am indebted to Centro Mexicano de Estudios en Farmacodependencia and to Dr. José Luis Díaz of the National University of Mexico who arranged and facilitated my studies.

Finally, I wish to express my special thanks to Catherine Delord of the University of Washington for translation of resource material.

SUGGESTED GENERAL READING

Brecher, Edward, et al, *Licit and Illicit Drugs*. Little, Brown, and Co., Boston, 1972.

Burgess, Anthony, *A Clockwork Orange*. Ballantine, New York, 1971.

Burroughs, William, *The Job*. Grove Press, New York, 1971.

Castaneda, Carlos, *A Separate Reality—Further Conversations with Don Juan*. Simon and Schuster, New York, 1971.

Castaneda, Carlos, *Journey to Ixtlan*. Simon and Schuster, New York, 1972.

Furst, Peter (ed.), *Flesh of the Gods: The Ritual Use of Hallucinogens*. Praeger, New York, 1972.

Gladwin, Thomas, *East is a Big Bird—Navigation and Logic on Puluwat Atoll*. Harvard Press, Cambridge, 1970.

Harner, Michael (ed.), *Hallucinogens and Shamanism*. Oxford Press, Oxford, 1973.

Heim, Roger, and Wasson, R. Gordon, *Les Champignons Hallucinogènes du Mexique*. Éditions du Muséum National d'Histoire Naturelle, Paris, 1958.

Joyce, James, *Ulysses*. Monarch Library, New York, 1961.

Laing, Ronald, *The Politics of Experience*. Pantheon, New York, 1967.

Lamb, F. Bruce, *Wizard of the Upper Amazon*. Houghton-Mifflin, Boston, 1975.

Lilly, John, *The Mind of the Dolphin*. Doubleday, New York, 1967.

Maslow, Abraham, *Toward a Psychology of Being*. Van Nostrand Reinhold, New York, 1968.

Miller, Orson, *Mushrooms of North America*. Dutton, New York, 1971.

Mortimer, W. Golden, *History of Coca: Divine Plant of the Incas*. And/Or Press, San Francisco, 1974.

Prescott, William, *The History of the Conquest of Mexico*. University of Chicago Press, Chicago, 1966.

Ram Dass, Baba, *Be Here Now*. Crown, New York, 1971.

Robbe-Grillet, Alain, *The Erasers*. Grove Press, New York, 1964.

Simeons, Albert, *Man's Presumptuous Brain: An Evolutionary Interpretation of Psychosomatic Diseases*. Dutton, New York, 1961.

Stapledon, Olaf, *Last and First Men*. Dover, New York, 1968.

Szasz, Thomas, *Ceremonial Chemistry: The Ritual Persecution of Drugs, Addicts, and Pushers*. Doubleday, New York, 1974.

Thoreau, Henry, *Walden, Or Life in the Woods*. Washington Square Press, New York, 1968.

Wasson, R. Gordon, *María Sabina and her Mazatec Mushroom Velada*. Harcourt, Brace, Jovanovich, New York, 1974.

Wasson, R. Gordon, *Soma: Divine Mushroom of Immortality*. Harcourt, Brace, and World, New York, 1968.

Wasson, Valentina P., and Wasson, R. Gordon, *Mushrooms, Russia, and History*. Pantheon Books, New York, 1957.

Weil, Andrew, *The Natural Mind—A New Way of Looking at Drugs and the Higher Consciousness*. Houghton-Mifflin, Boston, 1972.

Zinberg, Norman, and Robertson, John, *Drugs and the Public*. Simon and Schuster, New York, 1972.

BIBLIOGRAPHY

1. Aghajanian, G., et al, in *Psychotomimetic Drugs.* (Efron, D., ed.) pp 165–170, Raven Press, N.Y., 1970.
2. Aldrich, M., *A Brief Legal History of Marijuana.* Do It Now Foundation, Phoenix, 1971.
3. Aldrich, M., personal communications, 1/75–3/75.
4. Alexander, L., in *Origin and Mechanisms of Hallucinations.* (Keup, W., ed.), pp 133–148, Plenum Press, N.Y., 1970.
5. Allegro, J., *The Sacred Mushroom and the Cross.* Doubleday, N.Y., 1970.
6. Allen, T., and Argus, B., *Am. J. Psychiatry* 125: 632–637, 1968.
7. Anderson, E., *Am. J. Botany* 50: 724–732, 1963.
8. Avery, A., et al, *Blakeslee: The Genus Datura.* Ronato Press, N.Y., 1959.
9. Backer, C., *Flora of Java.* Nordhoff, Groningen, Netherlands, 1963.
10. Barrington, E., *The Chemical Basis of Physiological Regulation.* Scott, Foresman, and Co., Glenview, Ill., 1968.
11. Baxter, R., et al, *Nature* 185: 466–467, 1960.
12. Benedict, R., et al, *Lloydia* 25: 156–159, 1962.
13. Benedict, R., *Lloydia* 29: 333, 1966.
14. Berg, L., (ed.), *Description of Kamchatkaland.* p 236, Soviet Academy of Sciences, 1949.
15. Bloom, F., and Giarman, N., *Ann. Rev. Pharmacology* 8: 229–258, 1968.
16. Bose, B., et al, *Arch. Int. Pharmacodynam. Ther.* 147: 291, 1964.
17. Boydston, J., personal communications, 1/74–4/75.
18. Brady, L., and Benedict, R., *J. Pharm. Sci.* 61(2): 318, 1972.
19. Brady, L., and Tyler, V., *J. Am. Pharm. Assoc.* 48: 417, 1959.
20. Britton, M., and Rose, J., *The Cactaceae.* Carnegie Institute, Washington, 1922.
21. Brown, W., *Introduction to Organic and Biochemistry.* Willard Grant, Boston, 1972.
22. Bruce, P., *Economic History of Virginia in the 17th Century,* Vol. II. Peter Smith, N.Y., 1935.

23. Buck, R., *New Eng. J. Med.* 276: 391–392, 1967.

24. Burgess, A., *A Clockwork Orange.* Ballantine, N.Y., 1971.

25. Burroughs, W., *Junkie.* Ace Books, N.Y., 1953.

26. Burroughs, W., *Naked Lunch.* Grove Press, N.Y., 1963.

27. Castaneda, C., *The Teachings of Don Juan: A Yaqui Way of Knowledge.* Ballantine, N.Y., 1969.

28. Catalfomo, P., and Tyler, V., *J. Pharm. Sci.* 50: 687, 1961.

29. Cerletti, A., in *Neuropharmacology.* (Bradley, P., et al, eds.), pp 291–295, Elsevier, Amsterdam, 1959.

30. Cerletti, A., et al, *Schweiz. Apoth. Ztg.* 101: 210, 1963.

31. Chao, J., and DerMarderosian, A., *Phytochemistry* 12: 2435–2440, 1973.

32. Chilton, W., personal communications, 2/74–4/75.

33. Chilton, W., and Ott, J., *Lloydia,* 39(2&3): 150–157, 1976.

34. Clark, A., in *The Witch Cult in Western Europe.* (Murray, M.), p 279, Oxford University Press, London, 1921.

35. Claussen, U., and Korte, F., *Naturwissenschaften* 53: 541–546, 1966.

36. Cohen, S., *Ann. Rev. Pharmacology* 7: 301–318, 1967.

37. Cohen, S., *The Beyond Within.* Atheneum, N.Y., 1964.

38. Costa, E., *Proc. Soc. Exptl. Biol. Med.* 91: 39, 1956.

39. Coulter, J., *Preliminary Revision of the North American Species of Cactus, Anhalonium, and Lophophora.* Contributions from the U.S. Herbarium, Vol. 3, U.S. Government Printing Office, Washington, 1896.

40. Curtis, D., et al, *British J. Pharm.* 18: 217, 1962.

41. Dandiya, P., et al, *British J. Pharm.* 20(3): 436–442, 1963.

42. Dandiya, P., et al, *Indian J. Med. Research* 50: 46–60, 1962.

43. Dearness, J., *Mycologia* 27: 85–86, 1935.

44. deCarollis, S., et al, *Psychopharmacologia* 15: 186, 1969.

45. del Pozo, E., in *Ethnopharmacologic Search for Psychoactive Drugs.* (Efron, D., et al, eds.), pp 59–76, U.S. Public Health Service Publication #1645, Washington, 1967.

46. del Pozo, E., *Ann. Rev. Pharmacology* 6: 9–18, 1966.

47. DeQuincey, T., *Confessions of an English Opium Eater.* Signet Books, N.Y., 1966.

48. Dewey, D., et al, *Nature* (London) 226: 1265, 1970.

49. Díaz, J., personal communication, 7/74.

50. Dishotsky, N., et al, *Science* 172: 431, 1971.

51. Dittmar, C., *Historical Reports.* St. Petersburg, 1890.

52. Dobkin de Rios, M., *Current Anthro.* 15(2): 147, 1974.

53. Downing, D., *Psychopharmacological Agents.* Academic Press, N.Y., 1964.

54. Drake, W., *The Cultivator's Handbook of Marijuana*. Wingbow Press, Berkeley, Cal., 1970.

55. Dunn, E., *Current Anthro.* 14: 488–492, 1973.

56. Eccles, J., *The Physiology of Synapses*. Springer-Verlag, N.Y., 1964.

57. Efron, D., et al, (eds.), *Ethnopharmacologic Search for Psychoactive Drugs*. U.S. Public Health Service Publication #1645, Washington, 1967.

58. Eliade, M., *Shamanism: Archaic Techniques of Ecstasy*. Princeton University Press, Princeton, N.J., 1972.

59. El-Zoghby, S., et al, *Biochem. Pharmacol.* 19: 1661–1667, 1970.

60. Emboden, W., *Cannabis: A Polytypic Genus*. publication pending in *Economic Botany*.

61. Emboden, W., *Narcotic Plants*. MacMillan, N.Y., 1972.

62. Enos, L., *A Key to the American Psilocybin Mushroom*. Youniverse, Lemon Grove, Cal., 1970.

63. Eugster, C., in *Ethnopharmacologic Search for Psychoactive Drugs*. (Efron, D., et al, eds.), pp 416–418, U.S. Public Health Service Publication #1645, Washington, 1967.

64. Eugster, C., *Naturwissenschaften* 55: 305–313, 1968.

65. Eugster, C., *Tetrahedron Letters* 23:1813–1815, 1965.

66. Fischer, R., *Rev. Can. Biol.* 17: 389, 1958.

67. Flattery, D., personal communication, 1/75.

68. Frazier, J., *The Marijuana Farmers*. Solar Age Press, New Orleans, La., 1974.

69. Garattini, S., in *Hashish: Its Chemistry and Pharmacology*. Ciba Foundation Study Group #21, pp 70–82, Churchill, London, 1965.

70. Gershon, S., et al, *Arch. Inter. Pharmacodynamie* 135: 31–56, 1962.

71. Goodspeed, T., *The Genus Nicotiana*. Chronica Botanica, Waltham, Mass., 1954.

72. Guzmán, G., *Bol. Soc. Bot. Mex.* 24: 14–34, 1959.

73. Guzmán, G., *Bol. Soc. Mex. Mic.* 6: 43, 1972.

74. Guzmán, G., personal communications, 8/74–5/75.

75. Harlow, A., and Woolsey, C., *Biochemical and Biological Bases of Behavior*. University of Wisconsin Press, Madison, 1965.

76. Harner, M., (ed.), *Hallucinogens and Shamanism*. Oxford University Press, Oxford, 1973.

77. Harner, M., in *Hallucinogens and Shamanism*. (Harner, M., ed.), pp 125–150, Oxford University Press, Oxford, 1973.

78. Hatfield, G., et al, *J. Pharm. Sci.* 58: 1298, 1969.

79. Heim, R., and Wasson, R., *Les Champignons Hallucinogènes du Mexique*. Éditions du Muséum National d'Histoire Naturelle, Paris, 1958.

80. Heim, R., *Les Champignons Toxiques et Hallucinogènes.* N. Boubée, Paris, 1963.

81. Heim, R., *Nouvelles Investigations sur les Champignons Hallucinogènes.* Éditions du Muséum National d'Histoire Naturelle, Paris, 1967.

82. Heim, R., *Revue de Mycologie* 22: 183–198, 1957.

83. Helmer, J., and Victorisz, T., *Drug Use, The Labor Market, and Class Conflict.* Drug Abuse Council, Washington, 1974.

84. Hernández, F., *Nova Plantarum, Animalium, et Mineralium Mexicanorum Historia.* Rome, 1651.

85. Himwich, H., and Alpers, H., *Ann. Rev. Pharmacology* 10: 313–334, 1970.

86. Hitchcock, C., and Cronquist, A., *Vascular Plants of the Pacific Northwest.* University of Washington Press, Seattle, 1959.

87. Hodgkins, J., et al, *Tetrahedron Letters* 14: 1321–1324, 1967.

88. Hoffer, A., and Osmond, H., *The Hallucinogens.* Academic Press, N.Y., 1967.

89. Hofmann, A., *Bot. Mus. Leaf. Har. U.* 20: 194–212, 1963.

90. Hofmann, A., in *Drugs Affecting the Central Nervous System,* Vol. II. (Burger, A., ed.), pp 169–236, Decker, N.Y., 1968.

91. Hofmann, A., and Heim, R., *C. R. Acad. Sci.* 257: 10–12, 1963.

92. Holmberg, G., et al, *Psychopharmacologia* 2: 93, 1961.

93. Horita, A., et al, *Toxicol. Appl. Pharmacol.* 4: 730, 1962.

94. Hylin, J., and Watson, D., *Science* 148: 499–500, 1965.

95. Isbell, H., et al, *Psychopharmacologia* 11: 184–188, 1967.

96. Isbell, H., et al, *Psychopharmacologia* 2: 147, 1961.

97. Janiger, O., and Dobkin de Rios, M., *Med. Anthro. News.* IV, August, 1973.

98. Johnston, G., *Psychopharmacologia* 22: 230, 1971.

99. Kalberer, F., et al, *Biochem. Pharmacol.* 11: 261, 1962..

100. Katz, B., *Nerve, Muscle, and Synapse.* McGraw Hill, N.Y., 1966.

101. König-Bersin, P., et al, *Psychopharmacologia* 18: 1, 1970.

102. Kühner, R., and Romagnesi, A., *Lib. de l'Acad. Med.* Paris, 1953.

103. Landa, D., *Relación de las Cosas de Yucatán.* (Tozzer, A., trans.), Peabody Museum, Cambridge, Mass., 1941.

104. Lange, M., and Hora, F., *Mushrooms and Toadstools.* Dutton, N.Y., 1963.

105. Lehninger, A., *Biochemistry.* Worth, N.Y., 1970.

106. Lewin, L., *Arch. Exptl. Path. Pharmakol.* 129: 133–149, 1928.

107. Lewis, L., *Biochem. J.* 57: 680, 1954.

108. Lilly, J., *Center of the Cyclone.* Julian Press, N.Y., 1972.

109. Loewe, S., *J. Pharmacol. Exper. Ther.* 88: 154–161, 1946.

110. Loewe, S., in *The Marijuana Problem of the City of New York.* pp 149–212, Jaques Cattell Press, Lancaster, Pa., 1944.

111. Lowy, B., *Mycologia* 63: 983–993, 1971.

112. Lowy, B., *Mycologia* 64: 816–821, 1972.

113. Lowy, B., *Mycologia* 66: 188–191, 1974.

114. Lumholz, C., *Unknown Mexico,* Vol. I. Scribner, N.Y., 1902.

115. McLaughlin, J., personal communication, 10/74.

116. McLaughlin, J., *Lloydia* 32(3): 392–394, 1969.

117. Makeshwari, J., *The Flora of Delhi.* Council of Scientific and Industrial Research, New Delhi, 1961.

118. Martin, P., *Science* 179: 969–974, 1973.

119. Maugh, T., *Science* 185: 684, 1974.

120. Michelet, J., *Satanism and Witchcraft.* (Allinson, A., trans.), Citadel, N.Y., 1965.

121. Miller, O., *Mushrooms of North America.* Dutton, N.Y., 1971.

122. Miras, C., in *Hashish: Its Chemistry and Pharmacology.* Ciba Foundation Study Group #21, p 37, Churchill, London, 1965.

123. Munn, H., in *Hallucinogens and Shamanism.* (Harner, M., ed.), pp 86–122, Oxford University Press, Oxford, 1973.

124. Murphee, H., *Ann. N.Y. Acad. Sci.* 142: 1–333, 1967.

125. Nakamura, Y., *Nippon Yakurigaku Zasshi* 61: 42–68, 1965.

126. Neal, J., and McLaughlin, J., *Lloydia* 33(3): 395–396, 1970.

127. Nietzsche, F., *Thus Spake Zarathustra.* (Kaufman, W., trans.), Viking, N.Y., 1966.

128. Norquist, D., and McLaughlin, J., *J. Pharm. Sci.* 39(12): 1840–1841, 1970.

129. Ola'h, G., *Le Genre Panaeolus,* Revue de Mycologie Ser. 10, Muséum National d'Histoire Naturelle, Paris, 1969.

130. Ott, J., unpublished field and laboratory notes, 1974–1975.

131. Ott, J., et al, *Physiological Chemistry and Physics* 7: 381–384, 1975.

132. Ott, J., *J. Psyched. Drugs* 8(1): 27–35, 1976.

133. Parke, R., and Salisbury, F., *Vascular Plants: Form and Function.* Wadsworth, Belmont, Cal., 1969.

134. Patkanov, S., *The Irtysh Ostyak and their Folk-Poetry.* St. Petersburg, 1897.

135. Pennes, H., et al, *Am. J. Psychiatry* 133: 887–892, 1957.

136. Pletscher, A., *Helv. Physiol. Pharmacol. Acta* 17: 202–214, 1959.

137. Pletscher, A., in *Psychotropic Drugs.* p 468, Elsevier, Amsterdam, 1957.

138. Poindexter, E., and Carpenter, R., *Chemistry and Industry.* p 176, 1962.

139. Pollock, S., personal communication, 4/75.

140. Prain, D., *Bengal Plants,* Vol II. Botanical Survey of India, Calcutta, 1963.

141. Prescott, W., *The History of the Conquest of Mexico.* University of Chicago Press, Chicago, 1966.

142. Proust, M., *The Guermantes Way.* (Moncrieff, C., trans.), Random House, N.Y., 1970.

143. Puget Sound Mycological Society, *Mushroom Poisoning in the Pacific Northwest.* Seattle, 1972.

144. Ratcliffe, B., personal communications, 12/74-4/75.

145. Robbers, J., et al, *Lloydia* 32: 399-400, 1969.

146. Roderick, B., personal communications, 2/74-2/75.

147. Romagnesi, M., *Bol. Soc. Bot. France* 80: IV-V, 1964.

148. Safford, W., *J. Hered.* 6: 291-311, 1915.

149. Saleminck, C., *Planta Med.* 12: 397, 1964.

150. Schultes, R., *Am. Anthro.* 40: 698-715, 1938.

151. Schultes, R., *Am. Anthro.* 42: 429-443, 1940.

152. Schultes, R., in *Flesh of the Gods.* (Furst, P., ed.) pp 3-54, Praeger, N.Y., 1972.

153. Schultes, R., *Ann. Rev. Plant Physiol.* 21: 571-594, 1970.

154. Schultes, R., *Bot. Mus. Leaf. Har. U.* 1: 3-45, 1941.

155. Schultes, R., lecture at University of Washington, 3/1/74.

156. Schultes, R., *Lloydia* 29: 293-298, 1966.

157. Schultes, R., personal communications, 3/74-9/74.

158. Schultes, R., *Science* 163: 245-254, 1969.

159. Schultes, R., and Hofmann, A., *The Botany and Chemistry of Hallucinogens.* C. C. Thomas, Springfield, Ill., 1973.

160. Shafer, R., *Mycologia* 57: 318-319, 1965.

161. Sharma, J., et al, *Nature* 192: 1299-1300, 1961.

162. Shulgin, A., personal communication, 1/75.

163. Singer, R., *Mycologia* 50: 239-261, 1958.

164. Singer, R., and Smith, A., *Mycologia* 50: 262-303, 1958.

165. Silverstein, M., and Lessin, P., *Science* 186: 740-741, 1974.

166. Späth, E., *Monatsch. Chemic.* (Wien) 40: 129-154, 1919.

167. Stecher, G., (ed.), *Merck Index.* Merck Inc., Rathway, N.J., 1968.

168. Steller, G., *Description of Kamchatka, Its Inhabitants, Their Customs, Names, Way of Life, and Different Habits.* Leipzig, 1774.

169. Stoll, A., and Jucker, E., *Angewandte Chemie* 66: 376-386, 1954.

170. Stoll, A., et al, *Experientia* 11: 396, 1955.

171. Stoll, A., et al, *Helv. Chim. Acta* 37: 820, 1954.

172. Szasz, T., *Ceremonial Chemistry: The Ritual Persecution of Drugs, Addicts, and Pushers.* Doubleday, N.Y., 1974.

173. Tachibana, T., *Kobe Ika Daigaku Kiyo* 13: 649–657, 1958.

174. Talbot, E., and Vining, L., *Can. J. Bot.* 41: 639, 1963.

175. Theobald, W., et al, *Arzneimittelforsch.* 18: 311–315, 1968.

176. Tjio, J., et al, *J. Am. Med. Assoc.* 210: 849, 1969.

177. Turner, N., and Bell, H., *Econ. Bot.* 27: 257–310, 1973.

178. Tyler, V., *Lloydia* 24: 71–74, 1961.

179. Tyler, V., and Gröger, D., *Planta Med.* 12: 338, 1964.

180. Vergeer, P., personal communication, 4/75.

181. Walters, M., *Mycologia* 57: 837–838, 1965.

182. Walton, R., *Marijuana: America's New Drug Problem.* Lippincott, Philadelphia, 1938.

183. Waser, P., *Bull. Schweiz. Akad. Med. Wiss.* 27: 39, 1971.

184. Waser, P., in *Ethnopharmacologic Search for Psychoactive Drugs.* (Efron, D., et al, eds.), pp 419–439, U.S. Public Health Service Publication #1645, Washington, 1967.

185. Wasson, R., *Bot. Mus. Leaf. Har. U.* 20: 161–193, 1963.

186. Wasson, R., *Bot. Mus. Leaf. Har. U.* 23: 8, 1973.

187. Wasson, R., personal communications, 4/74–8/74.

188. Wasson, R., et al, *María Sabina and her Mazatec Mushroom Velada.* Harcourt, Brace, Jovanovich, N.Y., 1974.

189. Wasson, R., *Soma: Divine Mushroom of Immortality.* Harcourt, Brace, and World, N.Y., 1968.

190. Wasson, R., *Soma and the Fly Agaric: Mr. Wasson's Rejoinder to Professor Brough.* Botanical Museum of Harvard University, Cambridge, Mass., 1972.

191. Wasson, R., *Life* 42: 100–120, 1957.

192. Wasson, V., and Wasson, R., *Mushrooms, Russia, and History.* Pantheon Books, N.Y., 1957.

193. Watson, G., *Nutrition and your Mind.* Harper and Row, N.Y., 1972.

194. Weil, A., *The Natural Mind: A New Way of Looking at Drugs and the Higher Consciousness.* Houghton-Mifflin, N.Y., 1972.

195. Weil, A., personal communications, 10/74–4/75.

196. West, L., et al, *Lloydia* 37: 633, 1974.

197. Wiener, N., *Ann. Rev. Pharmacology* 10: 273–290, 1970.

198. Wilbert, J., in *Flesh of the Gods*. (Furst, P., ed.), pp 55–83, Praeger, N.Y., 1972.

199. Willaman, J., and Schubert, B., *Alkaloid Bearing Plants and their Contained Alkaloids*. U.S. Department of Agriculture Technical Bulletin #1234, Washington, 1961.

200. Wooley, N., et al, *Ann. N.Y. Acad. Sci.* 66: 649, 1957.

201. Wooley, D., et al, *Science* 119: 587, 1954.

202. Youngken, H., *Am. J. Pharmacy* July, 1924.

203. Youngken, H., *Am. J. Pharmacy* March, 1925.

204. Zinberg, N., *The Harvard Review* 1: 56–62, 1963.

205. Zuckerman, N., in *Origin and Mechanisms of Hallucinations*.(Keup, W., ed.), pp 133–148, Plenum Press, N.Y., 1970.

INDEX

A

acetyl choline, 107, 108, 113
acid freaks, 100
Acorus calamus, 39, 40, 114, 124
addicts, 99
Agaricaceae, 3
alcohol, xv, xvi
Algonquin Indians, 72
Allegro, John, 6, 97
Amanita species, 114–116, 125, 138
Amanita bisporigera, xiii, xviii
A. *cothurnata,* 6, 66
A. *gemmata,* 6
A. *muscaria,* 5, 6, 66, 85, 88–90, 115
A. *pantherina,* 6, 7, 8, 66, 115
A. *phalloides,* xiii, 138
A. *tenuifolia,* xiii
A. *verna,* xiii
A. *virosa,* xiii
α-amanitin, 138
amotivational syndrome, xvii
Anadenanthera species, 125
a-neg-la-kya, 70
Araceae, 37
Argyreia species, 109, 113, 123
Argyreia nervosa, 57, 58
Ariocarpus fissuratus, 43, 44
A. *retusus,* 44
Aryans, 6, 85, 89
asarone, 40, 114, 124
β-asarone, 40, 114
Atharva Veda, 52
Atropa species, xiii, 112, 127
Atropa belladonna, 65, 66, 70, 72

atropine, 6, 66, 68, 70, 72, 112, 115, 127
axons, 104–106, 111, 119
ayahuasca, 79
Aztec Indians, 16, 48, 60, 62, 68, 70, 72, 74, 92, 93

B

baby Hawaiian woodrose, 58
badoh, 62
badoh negro, 60
Banisteriopsis species, 74, 79, 109, 112, 126
Bardo thödol, 117
belladonna, 66
belladonna alkaloids, 70
bhangas, 52
bis-noryangonin, 113, 126
Blake, William, 102, 129
Buddhism, 90, 97, 117
Bufo species, 125
bufotenine, 6, 125
Burroughs, William, xv, xviii, 103

C

Cactaceae, 41, 109
Cannabaceae, 49
cannabinols, 52, 54
Cannabis species, xii, 68, 113, 126
Cannabis indica, 51, 52
C. *ruderalis,* 52
C. *sativa,* 51, 52, 53, 54

Hinduism, 90, 97, 98, 117
histamine, 107, 108
Hofmann, Albert, 62
holy water, 97
honguillos, xii
hordenine, 44
Huautla de Jiménez, 20
Huichol Indians, 48
humito, xii
Huxley, Aldous, 48
5-hydroxy tryptamine, 107, 108, 110–116, 123
hyoscyamine, 66, 68, 70, 72, 112, 127
Hyoscyamus species, 112, 127
hyperventilation, 118
hypnosis, 118

I

ibotenic acid, 6, 114, 115, 125
imipramine, 111
immune responses, marijuana and, xvii
Inquisition, 94, 95
Ipomoea species, 109, 113, 123
Ipomoea tricolor, 60
I. tuberosa, 58
I. violacea, 59
iproniazid, 111
iso-lysergic acid amide, 113

J

Jamestown weed, 72
Jimson weed, 72
junkies, 100

K

kava kava, 8, 113
Klüver, Heinrich, 48
Koryak Eskimos, 88
Kundalini yoga, 97

L

Landa, Diego de, 94
Leary, Timothy, 116, 117
Liberty Caps, 25, 27

Lophophora species, 113, 124
Lophophora williamsii, xii, 44, 45, 47, 48, 109
lophophorine, 48
lotus, 97
LSD, xviii, 62, 100, 109, 111, 113–118, 123
LSD-like effects, 58
LSD, sold as psilocybin, xviii
lysergic acid amide, 58, 60, 62, 109, 113, 123

M

macromerine, 45, 113, 124
MAO, 108, 110, 111, 112, 113
marijuana, 52, 54, 116
Marijuana Tax Act, 54
Mayan codices, 94
Mayan Indians, 62, 94, 97
Mazatec Indians, 18, 20, 23, 33, 35, 74, 82, 92
mbey san, 35
mescaline, 48, 109, 111, 113, 124
Mescalito, xii, 48
mescalotam, 48
methamphetamine, 109, 124
methanol, 138, 139
Miller, Orson, xiii
Mitchell, S. Weir, 48
Mixe Indians, 23
mujercitas, 31
muscarine, 6, 115
muscarinic poisonings, 66
muscazone, 6
muscimol, 6, 114, 115, 125
mushroom stones, 92
mycophobia, 86, 87, 93, 94

N

Nahuatl Indians, 23, 31, 92
nanacates, 92
Native American Church, 48
NE, 107, 108, 110, 111, 113, 115, 119, 124
neurohumors, 106–111
neurons, 104–108

niacin, 116, 118
Nicotiana species, 116, 126
Nicotiana rustica, 74
N. tabacum, 62, 73, 74, 75
nicotine, 74, 116, 126
Nietzsche, Friedrich, 121
N-methyl-3,4-dimethoxy-β
  phenethylamine, 44
N-methyl tyramine, 44
noradrenalin, 107
norepinephrine, 107, 124
norharman, 74, 116
nutritional pathology, 118, 119
Nymphaeaceae, 97

O

ololiuqui, 62
original sin, 96, 98
otherworld, 84–86, 88, 89, 96, 98

P

pajarito del monte, 33
Panaeolus species, xii, 109, 114, 123
Panaeolus campanulatus, 12
P. foenisecii, 9, 10
P. sphinctrinus, 11, 12
P. subbalteatus, 13, 14
pantothenic acid, 119
Patanjali, 90
Patkanov, Serafim K., 88
PCP, sold as psilocybin, xviii
Pearly Gates, 60
Peganum harmala, 74, 78, 79
Peganum species, 109, 112, 126
pentobarbitone, 114
peyoglutam, 48
peyote, xii, 44, 45, 48
peyote cimarrón, 44
peyotl, 16, 48
Pholiota species, 113, 126
picietl, 74
Piper methysticum, 8, 113
piule de barda, 35
Polyporus species, 113, 126
Proust, Marcel, xv

psilocin, 10, 12, 14, 16, 18, 20, 23, 25,
  27, 31, 33, 35, 109, 114, 115, 123
Psilocybe species, xii, 18, 20, 109, 114,
  123, 138
Psilocybe baeocystis, 15, 16
P. caerulescens, 17, 18, 20
P. cubensis, 19, 20, 21
P. cyanescens, 22, 23
P. mexicana, 114
P. muliercula, 30, 31
P. pelliculosa, 24, 25, 27
P. semilanceata, 25, 26, 27
P. stuntzii, 28, 29
P. wassonii, 30, 31
P. yungensis, 32, 33
P. zapotecorum, 34, 35
psilocybin, 10, 12, 14, 16, 18, 20, 23,
  25, 27, 29, 31, 33, 35, 111, 114–116,
  123
psilocybin, on illicit market, xviii
Pure Food and Drug Act, 54

Q

quauhyetl, 62, 74

R

Raleigh, Sir Walter, 74
Rauvolfia serpentina, 109
Rauvolfia species, 128
reserpine, 109, 110, 113, 128
Rg Veda, 6, 79, 85, 89, 90, 117, 135
Rivea species, 109, 113, 123
Rivea corymbosa, 61, 62, 74
Roquet, Salvador, 100
Roseocactus fissuratus, 43

S

Sabina, María, xi, xii
sahasrara chakra, 97
San Isidro, 20
Santa Claus, 97
Santo Niño de Atocha, 35
scopolamine, 66, 68, 70, 72, 112, 127
Scotch broom, 70
señoritas, 31